'The intersection of geography and the humanities provides fertile ground for the exploration of public art, and this collection does not disappoint. Through a diverse range of case studies, from Birmingham to Budapest, this book offers a dazzling summary of the state-of-the-art in critical geohumanities to document the different ways that public art is encountered, and contested, in particular contexts. Rather than treating public art as part of the backdrop against which urban life is played out, this book puts it front and central. Essential reading for anyone interested in the new scholarly interactions occurring at the point at which geography, art and place converge.'

– Phil Hubbard, Professor in Urban Studies, King's College London, UK

'By analysing specific public art projects through the frames of "power (the politics of encountering), affect (the embodied ways of encountering) and diversity (the inclusiveness of encountering)" this invitational anthology incorporates a multiplicity of disciplines – anthropology, art, art history, aesthetics, geography, history, heritage studies, performance and media studies, and sociology, among others – using a variety of productive methodologies. International in scope, its projects are located in many areas not typically covered in the existing literature: Birmingham (England), Budapest, London, Melbourne, Saskatchewan, Singapore, South Africa, Toronto, Vancouver, and Zimbabwe. This volume provides a range of essential tools for considering how public art has been and continues to be experienced by a multiplicity of audiences and is variously integrated into our actual and virtual spaces.'

– Harriet Senie, Professor of Art History, City College of New York, USA

'A welcome intervention by geohumanities scholars into the contemporary debates around art and participation. Simply removing art from the gallery, and collaborating with audiences is not an answer. This collection explores the challenges of exploring the intersections between "art as politics" and "politics as art"; and the struggle inherent in situating and making sense of places, things, people, or collectives of humans and non-humans.'

– Andy C. Pratt, Professor of Cultural Economy and Director, Centre for Culture & the Creative Industries, City University of London, UK

'Public art and its role as placemaker, identity-definer and political tool is a rich area of research and study. Zebracki and Palmer have curated an excellent collection of essays that give fresh perspective on these themes. With a focus on diversity of location and audience, of context and implementation, this is essential reading.'

– Andrew Shoben, Professor of Public Art and Computation, Goldsmiths, University of London, UK

Public Art Encounters

Public art is produced and 'lived' within multiple, interlaced and contested political, economic, social and cultural-symbolic spheres. This lively collection is a mix of academic and practice-based writings that scrutinise conventional claims on the inclusiveness of public art practice. Contributions examine how various social differences, across class, ethnicity, age, gender, religion, ability and literacy, shape encounters with public art within the ambits of the design, regeneration and everyday experiences of public spaces. The chapters richly draw on case studies from the Global North and South, providing comprehensive insights into the experiences of encountering public art via a variety of scales and realms.

This book advances critical insights of how socially practised public arts articulate and cultivate geographies of social difference through the themes of power (the politics of encountering), affect (the embodied ways of encountering), and diversity (the inclusiveness of encountering). It will appeal to scholars, students and practitioners of cultural geography, the visual arts, urban studies, political studies and anthropology.

Martin Zebracki is Lecturer in Critical Human Geography at the University of Leeds, UK. He has written and talked widely at the crossroads of public art, social engagement and (sexual) citizenship. Zebracki is the co-editor of *The Everyday Practice of Public Art: Art, Space, and Social Inclusion* (with Cameron Cartiere).

Joni M. Palmer is an Adjunct Associate Professor in the Department of Geography and Environmental Studies and the Community and Regional Planning Program at the University of New Mexico, USA. Her professional life has covered a blend of practice and academia across arts and cultural planning over the past 25 years.

Public Art Encounters

Art, Space and Identity

Edited by Martin Zebracki and Joni M. Palmer

Routledge
Taylor & Francis Group
LONDON AND NEW YORK

First published 2018 by Routledge

2 Park Square, Milton Park, Abingdon, Oxfordshire OX14 4RN
52 Vanderbilt Avenue, New York, NY 10017

Routledge is an imprint of the Taylor & Francis Group, an informa business

First issued in paperback 2019

British Library Cataloguing in Publication Data
A catalogue record for this book is available from the British Library

Library of Congress Cataloging in Publication Data
Names: Zebracki, Martin, 1984- editor. | Palmer, Joni M. editor.
Title: Public art encounters / edited by Martin Zebracki and Joni M. Palmer.
Description: Abingdon, Oxon : New York, NY : Routledge is an imprint of the Taylor & Francis Group, an Informa Business, [2018] | Includes bibliographical references and index.
Identifiers: LCCN 2017015001| ISBN 9781472468796 (hardback) | ISBN 9781315602837 (ebook)
Subjects: LCSH: Public art. | Art and society. | Public spaces—Social aspects.
Classification: LCC N8725 .P83 2017 | DDC 711/.57—dc23
LC record available at https://lccn.loc.gov/201701500

ISBN: 978-1-4724-6879-6 (hbk)
ISBN: 978-0-367-36210-2 (pbk)

Typeset in Times New Roman
by Swales & Willis Ltd, Exeter, Devon, UK

Contents

Illustrations

Figures

Tables

Contributors

Candice Boyd, PhD, is an artist-geographer with a background in clinical psychology. She completed a second PhD in Cultural Geography and the Creative Arts at the University of Melbourne in 2015 and is now a Senior Fellow of its School of Geography. Her interests are in the geographies of mental health, cultures of sense and movement, therapeutic spaces, experiences of rurality, and contemporary museum geographies. She is author of *Non-Representational Geographies of Therapeutic Art Making: Thinking Through Practice* (2017).

Ruth Catlow is an artist and curator working with emancipatory network cultures, practices and poetics. She is co-founding co-director of Furtherfield, promoting the need for society and culture to evolve participatory, critical, aesthetic and ethical responses in parallel, bringing together cultural and technological innovation. Recent commissions include Time Is Speeding Up (20-21 Visual Arts Centre); Sociality-machine (Tate); and Play the Web We Want (Southbank Centre). Catlow is named by the Foundation for P2P alternatives in their list of 100 Women Co-creating the P2P Society. Profiles: ruthcatlow.net, furtherfield.org.

Cameron Cartiere, BFA, MA, PhD, is an Associate Professor at Emily Carr University of Art + Design. She is the author of several works on public art including *The Everyday Practice of Public Art: Art, Space, and Social Inclusion*, co-edited with Martin Zebracki (Routledge, 2016). As part of her research on the sustainable effects of public art, Cartiere is collaborating with artists, writers, scientists and new media researchers to convert neglected greenways within British Columbia into native pollinator pastures using public art as the driving force for environmental renewal.

Paul Clements has published journal articles and books on the theoretical, representational and practical aspects of the arts and culture more generally, especially on public art, cultural resistance, and social/cultural inclusion and exclusion. He has worked creatively with excluded communities as an artist and musician and presently lectures part-time at Goldsmiths College. He has recently published a book with Routledge entitled *The Creative Underground: Art, Politics and Everyday Life* (2016).

Stephen Forcer is Senior Lecturer in French, and currently Head of the Department of Modern Languages at the University of Birmingham. His research publications include *Dada as Text, Thought and Theory* (Oxford: Legenda Research Monographs in French Studies, 2015), *Modernist Song: The Poetry of Tristan Tzara* (Oxford: Legenda, 2006), and with Emma Wagstaff a co-edited special issue of *Nottingham French Studies on the French Avant-garde* (2011). With Saskia Warren he co-organised the Birmingham Surrealist Laboratory (2014). He is an Editorial Board member of the journal *Dada/Surrealism*.

Aaron Franks is a researcher, writer, performer and geographer. He worked in professional theatre from 1996 to 2006 and has degrees in Drama (BFA, Alberta), Social Justice and Equity Studies (MA, Brock) and Human Geography (PhD, Glasgow). From 2013 to 2016 Aaron was a postdoctoral researcher with the Centre for Environmental Health Equity and the Centre for Indigenous Research Creation at Queen's University, and is currently a Mitacs-SSHRC Visiting Fellow in Indigenous Research and Reconciliation at the Social Sciences and Humanities Research Council in Ottawa, Canada.

Andrew Gorman-Murray is Associate Professor of Social Sciences (Geography and Urban Studies) at Western Sydney University. His research includes: gender, sexuality and space; homemaking and households; neighbourhoods and belonging; place-making and mobilities; emotional geographies and wellbeing; visual and material cultures. He co-edited *Material Geographies of Household Sustainability* (2011, with Ruth Lane); *Sexuality, Rurality, and Geography* (2013, with Barbara Pini and Lia Bryant); *Masculinities and Place* (2014, with Peter Hopkins); *Queering the Interior* (2017, with Matt Cook). He is co-editor of *Emotion, Space and Society* and on the editorial boards of *Gender, Place and Culture*, *Social and Cultural Geography* and *ACME*.

Pauline Guinard is Associate Professor in the Department of Geography at the Ecole normale supérieure (Paris, France) and member of the research units UMR LAVUE-Mosaïques / UMR IHMC (associate). Her work focuses on the relationships between cities and arts, especially in the South African context. She examines both the role of arts in the making of urban spaces and the representations of cities in arts, notably in the visual ones. In 2014 she published a book entitled *Johannesburg, l'art d'inventer une ville* [Johannesburg, the art of inventing a city] at the Presses universitaires de Rennes.

Ashley Guindon, BFA, MAA, is a creative practitioner with an interdisciplinary art practice. Guindon's research explores long-term commitment to communities, creating opportunities for emerging artists and developing innovative projects with sustainable environmental, social and cultural impact. Guindon is sessional faculty at Emily Carr University and a coordinator for projects in the public realm.

Brenden Harvey has a master's degree in Social Anthropology from Dalhousie University, Halifax, Canada. His research interests include urban green space,

Internet culture, neighbourhood development and social class. Since 2015 he has been working in the field of community improvement as a Social Policy Research Analyst for the City of Calgary.

Harriet Hawkins is Professor of GeoHumanities at Royal Holloway, University of London. Her work focuses on the geographies of art works and art worlds. Her recent publications include *Creativity: Live, Work, Create* (Routledge, 2016) and *For Creative Geographies* (Routledge, 2014). Creative practices and collaborations are key to her research and she has worked with a range of individuals and organisations including Tate Modern, Arts Catalyst, Furtherfield and Swiss Artists in Labs. She is currently an AHRC Leadership Fellow researching 'Creating Earth Futures' and is Co-Director of the Royal Holloway Centre for the GeoHumanities.

Audrey Kobayashi is a professor and Queen's Research Chair in the Department of Geography, Queen's University. Her work ranges across issues of racism, gender, immigration, labour, urban poverty, disability and human rights. She has worked closely with groups in the Downtown East Side of Vancouver for 40 years, and many of her publications are inspired by community activism.

Dr Jason Luger is a human geographer and lecturer in urban studies at the University of San Francisco and UC Berkeley, and is an assistant editor at the *Journal of Urban Cultural Studies*. His research explores the intersection of urban space, politics, art and activism in a comparative urban framing. He has previously published in *IJURR*, *City*, *M/C Journal*, and *Geoforum* in addition to several book chapters, and has a co-edited volume (with Dr Julie Ren), which is entitled *Art and the City* (Routledge, 2017).

Gwen MacGregor is a visual artist and cultural geographer working across the disciplines of installation, video, photography and geographical scholarship. She has work in the Art Gallery of Ontario and the Royal Bank Collections. MacGregor is represented by MKG127 in Toronto. She also collaborates with the collective Workparty. MacGregor has an Honours BA from York University, a Master's in Geography from the University of Toronto and is a PhD Candidate in Geography with a SSHRC doctoral scholarship. Her dissertation explores the relationship between constructions of nationhood and how this is represented and contested in contemporary art.

Jeff Masuda is Canada Research Chair in Environmental Health Equity and an Associate Professor in the School of Kinesiology and Health Studies at Queen's University. He is a social geographer whose work bridges environmental justice, human rights and health. For the past 10 years he has partnered closely with urban activists, artists and community leaders to expose and oppose the symbolic and human violence of rapid gentrification, dispossession, and displacement in places like Vancouver, Toronto, and Winnipeg, Canada.

Zodwa Nyoni is a poet and playwright (published by Bloomsbury) and holds an MA in Writing for Performance and Publication, University of Leeds.

She has performed her work (inter)nationally, including: British Museum, Ilkley Literature Festival, National Gallery (Zimbabwe), Nuyorican Poets Café (New York) and Historic Hampton House (Miami). She won the Channel 4 Playwright's Scheme and was Writer-in-Residence at West Yorkshire Playhouse (WYP), Leeds. Nyoni's plays include: *Nine Lives* (2014, Oron Mor & WYP), *Boi Boi is Dead* (2015, WYP), *Love Again* (2016, BBC Radio 3), *Weathered Estates* (2017, University of Hull/City of Culture) and *Ode to Leeds* (2017, WYP).

Joni M. Palmer, MLA, PhD, lives in Albuquerque, New Mexico. She is an Adjunct Associate Professor in the Department of Geography and Environmental Studies and the Community and Regional Planning Program at the University of New Mexico. Her professional life has been a blend of practice and academia over the past 25 years. Her professional practice includes: arts and cultural planning, community planning and civic engagement, and landscape architecture. Her research interests include: cultural planning, landscape studies, public art, vernacular creativities and visual representation.

Martha Radice is an urban anthropologist whose research interests include public space, public art and public culture; interethnic relations; and neighbourhoods, community and sociability. Her recent publications include articles about cosmopolitanism, conviviality, and the publicness of public art, and a book co-edited with Alexandrine Boudreault-Fournier on public art interventions in Canadian cities. She is an Associate Professor in the Department of Sociology and Social Anthropology at Dalhousie University, Halifax, Canada.

Saskia Warren is Lecturer in Human Geography, University of Manchester. Previously she held the position of Research Fellow, University of Birmingham, where she led the project, Birmingham Surrealist Laboratory. She completed her PhD in Cultural Geography at University of Sheffield (2012), and MA in Art Gallery and Museum Studies at University of Leeds (2008). Her research has been funded by Heritage Lottery Fund, Collections Trust, Arts Connect, Engineering and Physical Sciences Research Council, and Arts and Humanities Research Council. Along with Phil Jones, she is co-editor of the book *Creative Economies, Creative Communities: Rethinking Place, Policy, and Practice* (Routledge, 2015).

Martin Zebracki, BScHons, MSc, PhD, is Lecturer in Critical Human Geography and Equality and Inclusion Leader in the School of Geography, University of Leeds, and serves as Secretary of the Sexualities and Queer Research Group at the Royal Geographical Society with Institute of British Geographers. His research revolves around intersecting geographies of public art, social engagement and (sexual) citizenship. Zebracki's previous anthology is *The Everyday Practice of Public Art: Art, Space, and Social Inclusion* (edited with Cameron Cartiere, Routledge, 2016). His latest publications focus on emerging digitally networked spaces of public art and and their potentials for and limitations to socially inclusive engagement. Profile: www.zebracki.org.

Acknowledgements

This anthology is the result of concerted action by scholars with a sincere commitment to the study of public art. We have encountered most of the contributors in previous collaborations ensuing from international conference events over the past five years. We owe a debt of gratitude to their incisive contributions to this collection, which provide compelling perspectives across the geohumanities as well as geographical regions.

An earlier version of Paul Clements' chapter appeared in *Art & the Public Sphere*, volume 2, numbers 1–3 (2014). The chapter by Martin Zebracki appeared as an article in the *International Journal of Cultural Studies*, first published online on 9 May 2016.

Moreover, we are most grateful to the commissioning editor Katy Crossan and editorial assistant Priscilla Corbett for guiding us through the project, as well as our copy editors Rob Wilkinson and Deirdre Harris for dotting the i's and crossing the t's. Special thanks goes to all authors for the inspiring teamwork and their patience throughout this publication journey.

Martin Zebracki (Leeds, UK) and
Joni Palmer (Albuquerque, NM)

Introduction

1 On encountering public art

Martin Zebracki and Joni M. Palmer

This is an invited edited volume from public art scholars across the geohumanities. The compilation includes an amalgamation of conceptual and empirical analyses of public art encounters based on a mixture of compelling qualitative and quantitative methodologies. It offers geographical analytical perspectives across the Global North and Global South, a comparative dimension that has remained largely overlooked in public art studies to date. The analyses particularly fill a niche in public art studies by adopting a multi-scalar approach, in that they provide comprehensive insights into encountering public art across the intersecting levels of the body/intimate, home, city, suburban, rural, region, state, world, digital realm, and so on, within multiple and interlaced political, economic, social and cultural-symbolic spheres.

We (the editors, and the contributing authors) believe that in order to advance public art scholarship, further insights are needed into how the socio-spatial ramifications and practices of public art reveal, cultivate and may articulate (alternative) geographies of social difference. Building on the recent anthology *The Everyday Practice of Public Art* edited by Cartiere and Zebracki (2016), and inspired by our many peers in the field, we critically pursue the question 'To what extent does public art practice live up to recurring fundamental claims of inclusive involvement, or is public art practice complicit in contributing to lived realities of exclusion?' An understanding of the substance of public artwork and its fluid spaces for incidental everyday engagements involves a concern with public interest (*inter-esse*) and everyday socially practiced space (e.g. De Certeau 1984, Lefebvre 1996). It is our hope that this volume begins a more thorough examination of a particular knowledge gap: how various intersectional social differences across identity markers, including class (Chapters 3, 4 and 5), ethnicity (Chapters 7, 9 and 10), age (Chapter 8), gender (Chapter 2), religion (Chapter 11) and ability/literacy (Chapters 6 and 12) play a part in shaping encounters with public art in the ambits of the design, regeneration and everyday experiences and imaginations of public spaces. This anthology does not claim to be exhaustive. It presents a concise, balanced collection of academic and practice-based writings about public art by a carefully selected set of authors. The work includes new voices and readings, and aims to provoke new ways of thinking about public art encounters, as well as 'doing' and managing public art practice within a variety of geographical conditions.

Hence, this anthology revolves around multivalent stories of encounter. First of all, this book reflects encounters amongst contributors in different stages of career, intellectual and disciplinary journeys, creative practices, as well as social and geographical contexts. Over the last five years, we have encountered most of the contributors in collaborative contexts of, among others, special journal issues, book projects and international conference events.[1] As is highly common in the digital academe today, we have mainly engaged with the contributors as so-called 'e-mailing bodies', hence involving a process of cyborgian encountering.

Encounter as noun, and as dynamic condition, can be fundamentally defined as an unexpected assembly, within which there are challenges, conciliatory confrontations, surprising collaborations, and so on – which resonates with antagonistic philosophies (Bishop 2012, Kester 2011). The term also functions as verb and therefore as action with a questioning disposition: to meet unexpectedly, to undergo, to experience, as if it were a phenomenological endeavour (e.g. Crouch 2010). The unexpected nature of encounter as dynamic condition and critical action is frequently at odds with blueprinted public art practices that are anchored in the dominion of formal policies. Public art is a socially situated and contested process (Miles 1997), and this subject matter therefore, after Kwon (2004), requires site-specific questions, answers and engagements.

Formal readings of public art, despite ongoing polemics about its ontology (e.g. Hein 2006), commonly understand public art as creations that are commissioned and designated for freely physically or visually accessible public places, which are meant for broad public interest and funded with public tax money (Cartiere and Willis 2008). However, this basic conceptualisation can be challenged in multiple ways based on complexions of space, e.g. multi-layered offline and online worlds, and time, e.g. the permanence of material artwork vis-à-vis the ephemerality of performance-based artwork. Also, public art neither starts with the artists' idea nor ends upon its installation or execution. Put simply, encountering public art is done by all and has no 'use-by date'. Considering the myriad of everyday spontaneous and informal manifestations of art in public spaces, commissioning authorities and qualified artists are not necessarily the driving forces behind its existence and for mapping out the destiny, form and modes for engagements (e.g. Knight 2008, Senie and Webster 1992).

Beyond the remit of formal commissioning and implementation, there are vast opportunities for appropriating, making and exchanging public art. Therefore, as the contributions in this volume address with greater empirical detail, public art encounters also encompass stories of everyday *co-creation* with variedly perceived uses and misuses of public art (e.g. Lossau and Stevens 2015). Co-creation may involve mental appropriations (e.g. endorsement, rejection) and physical actions (e.g. tactile interactions, or even vandalising) (e.g. Whybrow 2010). Also, in digitally networked space, there seem to be infinite opportunities for encountering artworks by geographically dispersed publics in material-digital (i.e. hybrid) spaces – after the artwork's material lifetime, if there is any (e.g. Zebracki, forthcoming).

Co-creation could ambiguously reveal support and resistance to hegemonic conventions of commissioned artworks and provide bottom-up, unsolicited curatorship. This alternative condition can unsettle official frameworks of public art production and, accordingly, deconstruct conformist dichotomies between expert and layperson, public and private, legal and illegal, and so on (e.g. Young 2013). The contributions attend to how informal encounters take shape through the intersections of material and immaterial spaces of the everyday life; for example, on city streets and squares (Chapters 3 and 7), in neighbourhoods (Chapters 5, 9 and 11), in transport (Chapter 2), in temporary and pop-up event spaces (Chapters 6 and 9), in parks (Chapter 4), over (mobile) social media at home and in open public spaces (Chapter 12), and in the subcultural underground (Chapter 8).

This volume's geographically differentiated focus, alongside informal engagements in the realm of the 'unforeseen', allow an understanding of the various socio-spatial capacities for encountering public artwork, as well as how the intertwined multiplicities of public art, space and identity are constantly renegotiated. The participants in this anthology, each in a distinct fashion, interrogate the complexities and ambiguities intrinsic to the geographies of public art encounter. The perspectives, of academics and practitioners, are situated across the disciplines of human and cultural geography, art history, anthropology, sociology, and performance and media studies.

We have accordingly identified three broad themes which frame the concerted effort and resonate with current concerns throughout the spectrum of the geohumanities: (1) *power*, indicating the politics of encountering; (2) *affect*, seen as the embodied ways of encountering; and (3) *diversity*, alluding to socially embracing qualities and realities of encountering and, hence, the inclusiveness of public art practice. These themes are anything but restrictive or mutually exclusive; rather, we pragmatically use them to structure the sections of this compilation. In the chapter-wise elucidation hereinafter, we explain how these terms are useful to framing contemporary socially practised spaces of public art.

Part I: power

The four chapters in this section attend to how encountering public art is inextricably bound up with power dynamics that invariably define the relationships between artists, artworks and public spaces. First, in 'Subverting surveillance: power and incivility in public transit art' (Chapter 2), Martha Radice and Brenden Harvey adopt an anthropological approach to examining how a temporary media art installation problematised such dynamics. Insights are drawn from *Time Transit* (Kim Morgan 2006) in Regina, Saskatchewan (Canada). The artwork displayed real-time photos taken of bus stops along the route along with text messages sent by passengers to the installation headquarters to be viewed on screens on an operational public bus. The authors identify visibility, location, technology and participation as key parameters for understanding *Time Transit* as a work of counter-surveillance.

The analysis of the encounters at a symbolic level challenges the power apparatus that states and corporations deploy in order to infringe on privacy in everyday public spaces. The authors also analysed the messages sent to the installation to evidence how it created conditions for the subversion of the normal social order of public transit. The authors argue that by opening up an anonymous, ephemeral forum for self-expression, the installation allowed participants to breach prevailing norms of civility with text messages that could be read as sexist, racist or homophobic. One (unanticipated) effect of the installation enabled the articulation of unequal power relations, which symbolically exclude certain social groups from participation in everyday life. The authors conclude that a public artwork, interrogating power relations in/of society, was itself appropriated by members of the public in ways that disrupted power relations at the micro-level. This case study, as such, raises critical questions about situational social impacts of public art.

As alluded to earlier, informal public art practices might have the potential to disrupt elitist structures and identities and top-down-led place-making initiatives. In '"Awaken the Dragon": participatory art-making and the grassroots in authoritarian Singapore' (Chapter 3), Jason Luger examines how the spontaneous praxis of dragon kilns provides the means by which locals, engulfed in tense relationships with authoritative city-state agents, might empower themselves to enliven marginalised urban heritage. Singapore has undergone a cultural turn in which the publicly funded arts have come to play an important socio-political role, where some strides have been made in engaging the citizenry in everyday arts and culture.

Thus, the arts provide paths for rising dissent and critical movements in Singapore, challenging the authoritarian rule that has traditionally defined this city-state. Although some of these movements keep a low profile, the empirical discussion of bottom-up, participatory art-making provides a useful lens for interrogating a broader mix of cultural and political fault lines. The study also raises questions about how dominant power hierarchies play out in commissioning art, 'parachuting' artists, and the role of publics within a city-state context which is increasingly divided and distanced by such heavy-handed governmental actions. 'Awaken the Dragon' emerges not only as an example of the tangible formation of Singaporean identity through shared artistic practice, but also as a metaphor for the ways grassroots actions re-appropriate and counter-produce space and cultural identity.

Paul Clements, in 'The construction of post-communist ideologies and re-branding of Budapest: the case study of Statue Park Museum' (Chapter 4), investigates how audiences encounter ambivalent conditions of the existing tourist-tailored public art park on the outskirts of Budapest. The author argues that such ambivalences are encrypted in changing state powers and hegemonic cultural memories. Clements critically frames the site of decommissioned public statuary as a peripheral space between modernist and postmodernist discourses of the public sphere. Based on reception theories, the study reveals the complexity of representation and especially the extent to which ideology overwrites the free play of interpretation of cultural heritage. Clements argues that the public art park, in multiple ways, is not only a bellwether of changing geopolitical times but also highlights contrasting understandings of public art in Eastern and Central Europe.

Clements asserts that Statue Park's meaning has vacillated in response to ideological shifts and the enduring memory of the communist era, the 'silent revolution', as well as new realities of post-communist 'freedom'. Such ambiguities have encouraged different gazes and identifications by public audiences, whose encounters with the artworks in the park include feelings of anger, ambivalence and irony. The analysis specifically attends to the re-branding of a redeveloped city via conspicuous commercial interests. Furthermore this is a city that has witnessed an ideological re-inscription of values of national and civic identity, as well as the development of an experience economy for budget airline tourists. Clements concludes that the contemporary representation of Statue Park corresponds to power dynamics in the political economy, where the public sphere has been significantly infiltrated by a neoliberal ideology which has largely 'disencountered' the communist past.

This section concludes with 'Sustainable influences of public art: a view on cultural capital and environmental impact' (Chapter 5), in which Cameron Cartiere and Ashley Guindon offer a situated methodological approach. They consider how we could gauge the effects of the politics of encountering: how is public art made, by and for whom is it made, and what are the intended and experienced benefits? On the basis of an in-depth case study on the research project *chART: Public Art Marpole* in Vancouver, British Columbia, the authors critically examine assumptions about the long-term social impacts of public art on the urban environment and its communities. They discuss the complexities involved in pursuing coordinated, comparative research strategies in examining what public art 'does' to people and places.

Cartiere and Guindon point out how the lack of both formal evaluation tools and clear understandings of targeted publics continue to challenge perceived rationales for public art, especially in the context of the Marpole neighbourhood which is increasingly facing significant economic, social and environmental challenges. The authors argue that public art is arguably one of the most contentious disciplines an artist could undertake; the field is especially complicated by the many stakeholders involved in urban (public art-led) regeneration policies. The case-study analysis shows that by building on smaller, temporary, and 'lower-risk' public art projects (22 in total in this case) the research project team was able to determine incremental qualitative and quantitative impacts, and discern growing impact over time (through trial and error). The authors reveal how such impacts can be concurrently tangible and embedded in memories of the artwork. Additionally, this study indicates how building recognition of and partnerships with community stakeholders – while making evaluation methods an integral part of the public art process – proves to be conducive to supporting power 'from below', thereby enhancing public ownership of public art.

Part II: affect

The three chapters in this section examine public art encounters at the embodied level, where non-representational understandings of experiencing and hence of the co-creation of public art are developed. Although non-representational theory has increasingly gained currency across the geohumanities (e.g. Anderson and Harrison 2010), its implications for grasping the social practices of public art at the intersections of art matter, place, and identity have, so far, remained under-explored.

The corporal capacities to sense and make sense of public art, in other words *affect*, are dealt with through the three chapters' foci on, respectively, how subjects are made and connected in participatory public (digital) arts (Chapter 6); how temporary site-specific public art practice may destabilise conventional notions of materiality and take on more-than-human forms (Chapter 7); and how young subjects embody freedoms of creative expressions through the geographical periphery of underground art (Chapter 8).

Harriet Hawkins and Ruth Catlow, in 'Shaping subjects, connecting communities, imagining futures? Critically investigating *Play Your Place*' (Chapter 6), impart that public art rhetoric has increasingly suggested claims to the promising possibilities of participatory arts practices for catalysing social relations. The authors ask researchers and practitioners to critically reflect on what it is that public art encounters actually 'do' in/for the world, an imperative recognised by a significant body of art theory. The authors develop an ethnographic account of a single project, and provide evidence for some new claims. Hawkins and Catlow empirically explore how the London-based *Play Your Place*, an interactive user-generated platform game, to which web users can contribute, functioned to shape subjects and connect communities through collective imaginations of new place-based futures. Through discussions about drawing and mapping, the authors identify the opportunities and limitations of the project. Also, future directions for examining forms and types of public art encounters are sketched along the nexus of art matter, body and technology.

In 'The production of temporary public space: site-specific installation and "vital materialities"' (Chapter 7), Gwen MacGregor engages with the notion of 'vital materiality' (Bennett 2010), and cognate 'new materiality' scholarship, to argue how embodied encounters with public art produce more-than-human particularities, thus challenging essentialist categorisations of subject and matter. The ephemeral scope of two site-specific artworks at the annual one-night Nuit Blanche visual arts festival in Toronto is used to demonstrate the fluidities between artist, artwork and space while disentangling the human-material ecologies of public art encounters. The author argues that the discussion about artworks in the festival has progressively relied on their framing as spectacle for large public audiences.

In particular, MacGregor challenges the understanding of humans as unique, or even as masters of nature, and instead analyses how public art practice reveals material agency, or a sense of affectivity in the non-human world. Public art encounters, therefore, are mediated through vibrant objects which are interconnected with both human and more-than-human activities, and beyond the categorisation of spectacle. So, this contribution challenges hegemonic notions of social identities as produced by public art practice via problematising 'new materiality' literature for its limited attention to more-than-human interactions. This literature is contrasted with black and indigenous scholarship, while acknowledging critical heterogeneous narratives of colonisation processes in North America, MacGregor parses how specificities of histories and places need to be recognised and articulated when both making and viewing a public artwork. The author suggests that we need to direct our attention to the multiple simultaneous encounters with artworks, which 'new materiality' alludes to but has not fully acknowledged or investigated as yet.

This section wraps up with '"All your drains belong to us": young people and the non-representational geographies of public art in drain tunnels' (Chapter 8) by Candice Boyd, who descends to the peripheral realm of underground graffiti in Melbourne. The author presents an argument in support of the 'therapeutic-ness' of underground art, where the focus on embodied encounters is identified as an understudied theme and young people as understudied age cohort in public art research. Boyds' theory- and empirical-driven analysis draws on a Deleuzian-inspired ontology, valorising the affirmative in graffiti, as well as urban play, whilst considering the ways in which underground artists are folded back on their creative practice contexts. The author argues that 'my art for my sake' is central to this kind of therapeutic art practice. However, Boyd critically indicates how social geographies of inclusion and exclusion are at play and serve to reinforce subjectivities. The author reflects on the ways in which the 'drain' represents a geographical scale, which affords young people particular freedoms of creative expression in developing their own identity. Along this particular ethico-aesthetic activity, Boyd's contribution highlights how more profound understandings of public art encounters can be elicited by paying attention to embodied experiences in youth-oriented spaces.

Part III: diversity

The contributions in this section coincide with a key concern with diversity, and ask the question 'How and to what degree does public art practice address social diversity and incite inclusive participation, and how can public art encounters be critically understood accordingly?' Societies are subject to increasing social mobilities and have become multicultural crucibles informing much public artwork that often endeavours to interconnect different places and heterogeneous communities. This section critically attends to this aim and the extent to which multifaceted place and social identities are (equally) addressed in public art's spaces of production, participation and co-creation.

In 'Mobilising the "Right to Remain" in Vancouver's Paueru-gai: an art-based participatory research intervention' (Chapter 9), Aaron Franks, Jeff Masuda, Audrey Kobayashi and the Right to Remain Community Fair Team investigate the implications for social inclusiveness of rapid gentrification currently sweeping the Downtown Eastside (DTES), a low-income neighbourhood in Vancouver, Canada. The Right to Remain Community Fair ('The Fair'), an art-based participatory research intervention, was established precisely as an attempt to form an antipode to this hyper-gentrification driven by neoliberal forces of urban change. Indigenous, Japanese, Chinese and African communities had historically constituted the formerly economically deprived DTES. The authors examine how 'The Fair', which ran from

2014–2016 as part of the multi-year project 'Revitalizing Japantown?', used three general strategies: (1) provoking with artistic *form*; (2) contesting and *spatialsing*; and (3) enabling *dialogue* within creative sites.

The authors demonstrate that the complex outcome was informed by a variety of artistic media addressing the Right to Remain alongside qualities of place which were conceived of as the basis of the diverse local community. These artistic media also explicitly challenged claims to territory and hegemonic forces, which had led to the dispossession of indigenous, marginalised, and racialised communities. This contribution highlights how public art practice can produce righteous alternatives to gentrification, which harbour (rather than efface) encounters with diversity.

Pauline Guinard takes us to the Global South in 'The art of (re)crossing the border: the *Border Farm* project in Maroi, South Africa' (Chapter 10). The author analyses public art encounters in the context of a migrant community in the culturally super-diverse border region of South Africa and Zimbabwe. Here, thousands of people from neighbouring and other African countries cross the border, either legally or illegally. After the contested re-election of President Mugabe, the influx of Zimbabweans drastically increased. Intrigued, if not alarmed, by the migrants' situation in the border region concerned, the author discusses how several artists attempted to understand and document the broader social and political conditions and the everyday lives of the people involved.

Based on an in-depth empirical analysis of the *Border Farm* project, initiated by Thenjiwe Nkosi in 2009, Guinard interrogates the artist's motivation to intervene. The author discusses how the public art project aimed to bridge places and identities of the people, who are divided through the migration process. Nkosi chose not to only involve people working on the farm in the making of the project. The artist genuinely endeavoured to build the project with the diverse communities. This resulted in the co-creation of a film which recounted the crossings, memories and traumas of the farm workers, whether having an illegal status or not. The film as such raises awareness around inclusionary as well as precarious exclusionary aspects of migration. Guinard's contribution questions the impacts of public art practice along both the material and immaterial borders of diverse people's everyday lives. This work uncovers how people and places are brought closer across social and cultural differences. It especially sheds new light on encountering public art at the crossroads of rural, border and migration spaces and underscores the empowering role of public art practice beyond the dominant sites of public art production.

In 'The Birmingham Surrealist Laboratory: unlocking community and the avant-garde in a super-diverse city' (Chapter 11), Saskia Warren and Stephen Forcer examine the opportunities of new digital facilities for encountering public artwork. They unlock complex issues of art, community and cultural sensitivity in the context of the super-diverse city of Birmingham. The authors discuss how the neighbourhood of Balsall Heath hosted the Birmingham Surrealist Group during the 1930s and 1950s, which was arguably the locale for Orthodox British Surrealism nationally. Today, Balsall Heath has been identified as falling within the lowest 5 per cent of neighbourhoods for multiple deprivations, and is home to a majority-minority community associated with conservative Islam. The neighbourhood offers productive ground for working through how digital technologies can be used to investigate multiple barriers to, and possibilities for, both mainstream and subversive manifestations of art, culture and heritage in an area with a conflicted past and high disengagement with institutions.

The analysis shows how the Birmingham Surrealist Laboratory project works directly with minority communities to foster multi-disciplinary and cross-sectoral dialogue, which may have the potential to provide significant insights for cultural and urban scholarship, arts practices and cultural policy. Using multi-touch digital facilities at the Digital Humanities Hub, University of Birmingham, the authors tested the ways in which Surrealism generated and nurtured connections with community participants recruited from Balsall Heath. In so doing, this project aimed to gain an understanding of how the often irreligious and overtly sexual and violent symbolism of Surrealist art is encountered by community groups from diverse ethnic backgrounds and religious traditions. This study locates experiences of avant-garde art and its legacy within a local landscape of fast-changing social, material and religious relations, which reshape individual agencies through collaborative public art encounters.

Contexts of the digital society and digital literacy are also key to this section's final study. In 'A cybergeography of public art encounter: the case of *Rubber Duck*' (Chapter 12), Martin Zebracki conveys that public art scholarship has largely been conducted on encounters in the offline world. Although virtual dimensions have become increasingly incorporated into public art practice, they have largely not been invested with equal attention. Zebracki addresses this research niche through an in-depth case study on the travelling *Rubber Duck* exhibition by Florentijn Hofman, which has been ongoing since 2012. The author scrutinises how publics identify with this public artwork and how they digitally mediate user-created content (which is mainly text- and image-based) through the various digitally networked spaces of interactive social media, online forums and news platforms (i.e. Web 2.0, or the post-Internet).

Zebracki develops the notion of 'virtual relationality' in the context of post-Internet communications. This notion implies the communication, (re-)negotiation and (re-)siting of public art's roles and meanings through social mediations within hybrid, material-digital contexts. Public art is examined along fluid cybergeographical understandings of its social and spatial publicness, temporalities and uses, which thus deconstruct conventional binaries such as material/digital, online/offline, permanence/ephemerality and human/non-human. Importantly, the study remains wary of digital illiteracies and the limitations of digitally networked spaces for inclusive participation, and encourages further research on the ethics of enlisting publics in digitally mediated public art practices and employing user-created content. In 'Everyday lived experience', as poetic encore appended to this book, Zodwa Nyoni responds to Zebracki's study on encountering digitally networked public art. This transformative response piece has been composed through the found poetry method and its provenance is explained in the appendix.

With this anthology, we and the contributing team endeavour to offer scholars and practitioners insights into how public art can be employed as an heuristic device for examining encounters with art matter, place and identity. We concurrently demonstrate how encounters with public art may shape both material and immaterial worlds, places and identity.

In the invited Afterword (Chapter 13), Andrew Gorman-Murray, through an artist-geographer's lens, incisively reflects on the trialectics between the practices, knowledges and experiences of public art. The author's synthesis acknowledges the complexity of the notions, methods and results of the thought-provoking studies. Gorman-Murray recognises that this collection constitutes a treasure-house for understanding public art theory and practice along the anthology's intersecting anchor themes of power, affect and diversity. Nonetheless, he suggests the themes of *mobile digital* public art, *rural* public art (beyond the dominant locus of the city; e.g. Florida 2012) and *multi-scalar* public art as research agendas for elaborating and thickening the geographies of public art encounter. Accordingly, we welcome peers to further canvass the spatio-temporal realms of encountering public art.

Note

1 See www.zebracki.org/conference-contributions/.

References

Anderson, B. and Harrison, P. 2010. *Taking-place: Nonrepresentational Theories and Geography*. Aldershot: Ashgate.

Bennett, J. 2010. *Vibrant Matter: A Political Ecology of Things*. Durham, NC: Duke University Press.

Bishop, C. 2012. *Artificial Hells: Participatory Art and the Politics of Spectatorship*. London: Verso.

Cartiere, C. and Willis, S. (eds) 2008. *The Practice of Public Art*. New York: Routledge.

Cartiere, C. and Zebracki, M. (eds) 2016. *The Everyday Practice of Public Art: Art, Space and Social Inclusion*. London: Routledge.

Crouch, D. 2010. *Flirting with Space: Journeys and Creativity*. Farnham: Ashgate.

De Certeau, M. 1984. *The Practice of Everyday Life*. Trans. S. Rendall. Berkeley, CA: University of California Press.

Florida, R. 2012. *The Rise of the Creative Class, Revisited*. New York: Basic Book.

Hein, H. 2006. *Public Art: Thinking Museum Differently*. New York: AltaMira.

Kester, G. 2011. *The One and the Many: Contemporary Collaborative Art in a Global Context*. Durham, NC: Duke University Press.

Knight, C. 2008. *Public Art: Theory, Practice and Populism*. Malden, MA: Blackwell.

Kwon, M. 2004. *One Place after Another: Site-specific Art and Locational Identity*. Cambridge, MA: MIT Press.

Lefebvre, H. 1996. *Writings on Cities*. Trans. and ed. E. Kofman and E. Lebas. Oxford: Blackwell.

Lossau, J. and Stevens, Q. 2015. *The Uses of Art in Public Space*. London: Routledge.

Miles, M. 1997. *Art, Space and the City: Public Art and Urban Futures*. London: Routledge.

Senie, H. and Webster, S. (eds) 1992. *Critical Issues in Public Art: Content, Context, and Controversy*. Washington, DC: Smithsonian Books.

Whybrow, N. 2010. *Art and the City*. London: I.B. Tauris.

Young, A. 2013. *Street Art, Public City*. New York: Routledge.

Zebracki, M. Forthcoming. Queerying public art in digitally networked space. *ACME: An International Journal for Critical Geographies*.

Part I

Power

2 Subverting surveillance

Power and incivility in public transit art

Martha Radice and Brenden Harvey

What happens when an art installation disrupts the social order of an urban public space? This chapter explores the power dynamics at work in *Time Transit*, an interactive media art installation by Kim Morgan that ran on a public bus in Regina, Saskatchewan (Canada) for two months in autumn 2006. *Time Transit* was created during Morgan's artist's residency at TRLabs, a research consortium associated with the University of Regina. It was installed on a bus operating on Route 4, which was fitted with two pairs of flat-screen monitors (one at the front and one toward the rear), a GPS system, a computer and a wireless network connection. Six digital cameras attached to computers transmitted photographs of six major bus stops along the route (Figure 2.1), and a website was set up for remote viewing and participation in the installation (www.timetransit.com). The connections between the hardware components were managed by a server at TRLabs. On the bus, the left-hand monitors showed translucent, overlaid photos of stops where the bus had already been – a layering of 'memories' – while the right-hand screens showed real-time single photos of where the bus was going – the stops ahead. The website showed a map of the route with the camera locations and real-time images and the real-time location of the bus. The key interactive element was that members of the public could send short text messages to the installation via mobile phone or the website. These scrolled along the screens beneath the photographs (Figure 2.2); if no new messages were incoming, the screens displayed messages selected at random from the pool of those already sent. A disclosure statement was posted on the bus and along the route to tell people about the project (Box 2.1). *Time Transit* was a complex, innovative assemblage of devices, software, data and people; its existence and functioning depended not only on close collaboration between Morgan and Craig Gelowitz, lead engineer on the project, but also on negotiation with institutions such as the City of Regina, Transit Regina and SaskTel, the regional telecommunications company.[1]

Time Transit constitutes a rich case study for analysing how different forms of power can circulate through public art. Unusually for an interactive installation, it left behind traces that are available for empirical analysis. While the photographs had to be discarded within 24 hours to comply with privacy legislation, the 3,960 text messages sent to the installation, which did not include any personally identifying data, were retained. As anthropologists who study the social and cultural

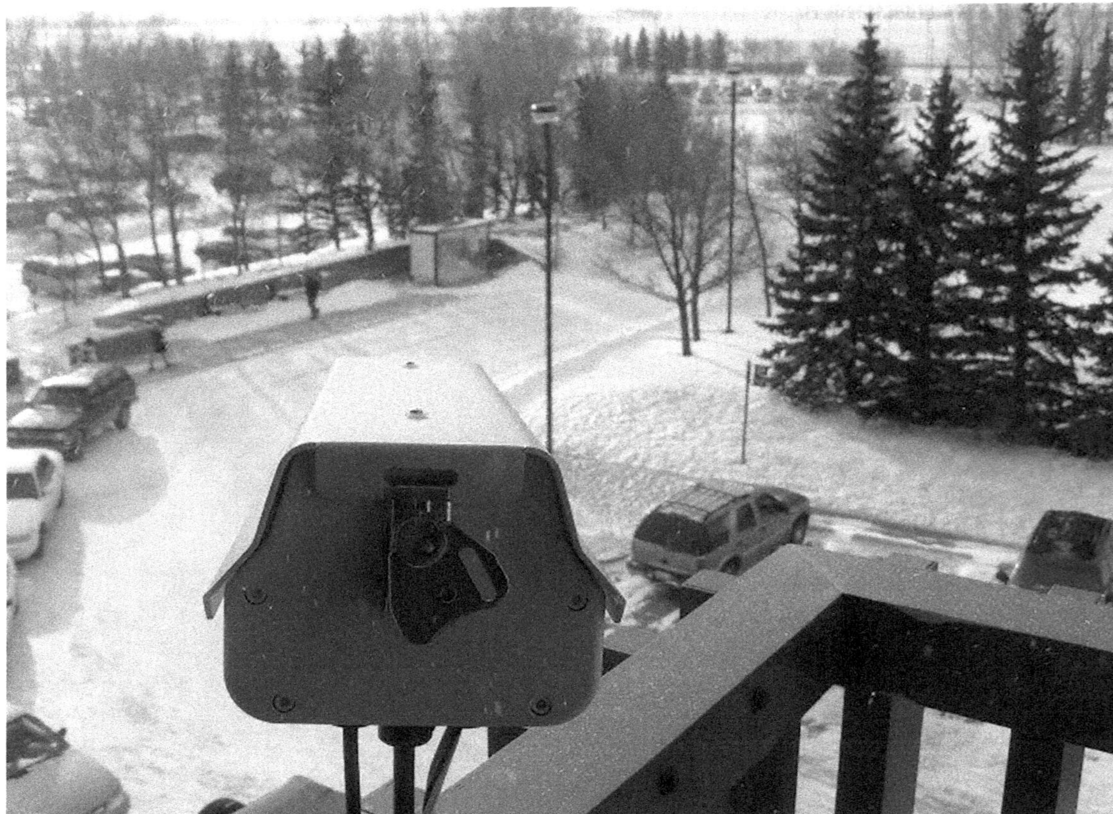

Figure 2.1 *Time Transit,* Regina, 2006: one of the cameras. Source: Kim Morgan.

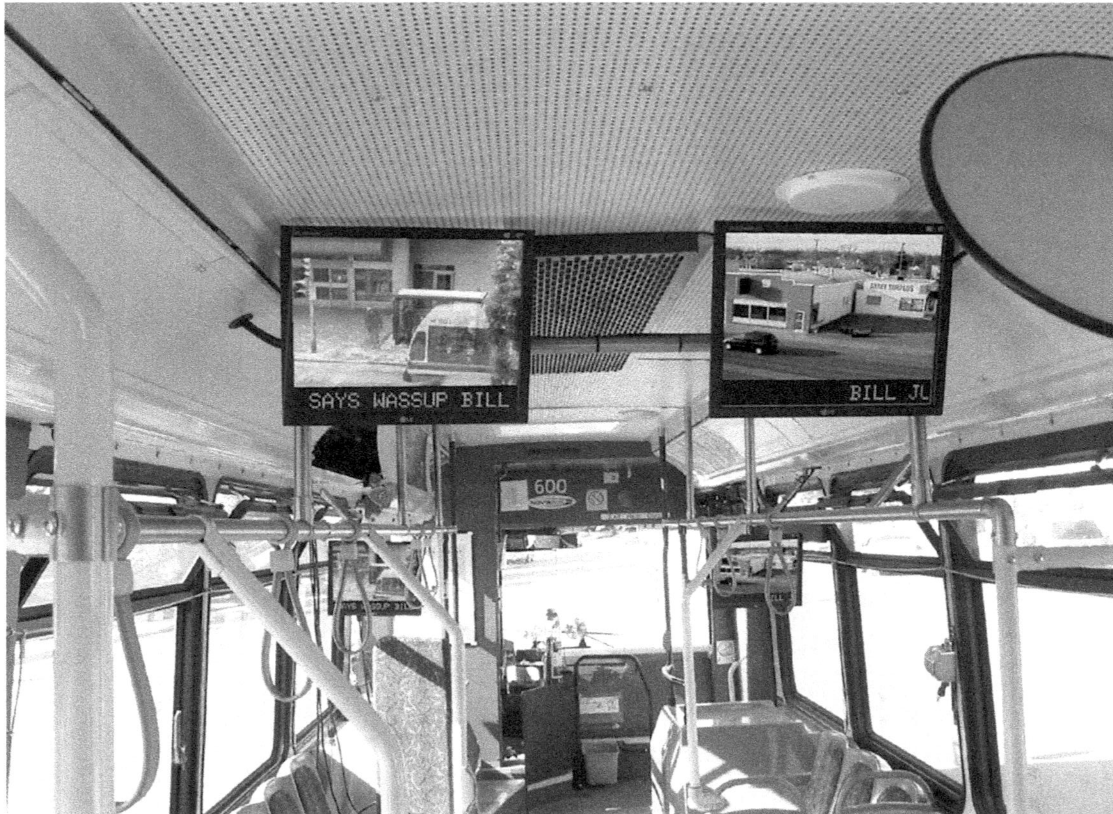

Figure 2.2 *Time Transit*, Regina, 2006: the view inside the bus. Source: Kim Morgan.

dynamics of urban space, we seek to understand how public art intervenes in the urban public – the routine or unpredictable configurations of people, places, information and events that make up a city's public life – and vice versa. The *Time Transit* text messages provide a rare opportunity to investigate one facet of how members of the public actually interacted with urban public art. We have conducted a qualitative analysis of this dataset, coding the messages along themes of content, genre and form. *Time Transit* deliberately engaged with the topic and technologies of electronic surveillance; it also created a space for subverting surveillance by enabling the public to 'talk back' in the installation. Our analysis shows that in so doing, it unsettled the social order of the public bus.

This chapter follows two lines of inquiry. First, we analyse *Time Transit* as a work of counter-surveillance. Specifically, we draw on sociological studies of

Box 2.1 Text of signage announcing *Time Transit*

Disclosure Statement

One of the buses traveling route **#4 (Walsh Acres/Hillside)** and the following bus stops are part of a mobile and interactive art installation entitled **Time Transit**.

11th Avenue/Cornwall Street
12th Avenue/Scarth Street
Albert Street/6th Avenue
Rochdale Boulevard/Radway Street
SIAST Campus Road/SIAST Main Campus
University Drive West/Riddell Centre

The project involves the use of surveillance **cameras**, a **Global Positioning System (GPS),** flat screen **monitors**, a public **website** and an **instant text messaging device**. Real-time images of individuals and activities at and within the vicinity of the bus/bus stops/bus shelters, and messages received from the public, **can be recorded and displayed on-board and on the website**, as part of the project.

For more information, call the City of Regina at 777-7000, or visit www.regina.ca and www.reginatransit.com.

This ground-breaking project encourages discussion about sustainable public transportation, and explores the relationship between art, engineering, public service and technology. It was designed and produced by TRLabs.

www.timetransit.com
text message: 1-306-526-6038

Source: Kim Morgan.

surveillance and 'artveillance' to identify four key concepts – visibility, location, technology and participation – which, along with our analysis of interviews that we conducted with Morgan and Gelowitz, their writing about the work (Morgan, Gelowitz and Benedicenti 2008, www.timetransit.com) and its press coverage (Anderson 2006, Beatty 2006, TRLabs 2006), illuminate how *Time Transit* challenged the 'institutional power asymmetries' embedded in surveillance systems (Monahan 2006: 516). Indeed, some such challenges were made explicit in a number of messages in the dataset. Second, we show how the text messages sent to *Time Transit* disrupted the ambience of urban civility that normally reigns on public transport. Although the bus company demanded a degree of censorship of the messages, some got through that were quite offensive – sexist, racist or homophobic. Using concepts from urban sociology and gender studies, we show how the unusual technological conditions of production of these messages (anonymity, mediation, ephemerality), combined with generalized hegemonic masculinity, allowed certain *Time Transit* contributors to breach the normal social order of the bus. We conclude by considering what implications *Time Transit*, as an artwork that enabled socially controversial interactivity, has for public art.

Art as counter-surveillance

One of the key roles of art is to trouble received wisdom, to induce uncertainty or wonderment about significant social and cultural phenomena (Barone and Eisner 2012). The burgeoning of surveillance, which raises questions about power relations, human rights, and the right to privacy versus the public good, is one such critical topic. Bennett et al. (2014: 6) define surveillance as 'any systematic focus on personal information in order to influence, manage, entitle, or control those whose information is collected'. Systems of surveillance have been expanding and intensifying for the last three or four decades, especially since the destruction wrought by hijacked planes on 11 September 2001. This expansion is driven by constant technological improvements, the profitability of the personal-information economy, and the entanglement of the neoliberal state with private lives and private enterprises; it is also bound up with increasing societal emphasis on risk management and national security (Bennett et al. 2014, Lyon 2003). While older surveillance methods involved personally monitoring individuals and manually compiling files on them that only select people could read, surveillance is now ubiquitous, largely automated and relatively accessible. Countless closed-circuit television cameras scan specific locations, but the biggest increase is in 'dataveillance' – surveillance extrapolated from the huge banks of personal data that are generated whenever humans interact with networked computer systems – surfing the Internet, buying things with credit or debit cards or using mobile phones, for instance (Lyon 2002). Such everyday surveillance contributes significantly to 'social sorting', meaning that data is abstracted from embodied persons to create sets of information and construct profiles that 'classify and monitor sets of people deemed risky and, sometimes, to exclude them from full participation in society' (Bennett et al. 2014: 47).

Enough artists have created work on surveillance – mainly visual, but also data-based – for social scientists to have started analysing it (Albrechtslund and Dubbeld 2005, Barnard-Wills and Barnard-Wills 2012, Brighenti 2010, Genosko 2006, McKay 2013, Monahan 2006). To delineate the topic, Brighenti has coined the broad term 'artveillance', 'the domain of the reciprocal influences and exchanges between art and surveillance', with its subset 'surveillance art' being 'every contemporary artwork that in some way hints to or deals with topics, concerns and procedures that falls within the interest of surveillance studies' (2010: 175). Drawing on surveillance studies, artveillance and social science at large, we identify four intertwined concepts that shed light on surveillance art in general and *Time Transit* in particular: visibility, location, technology and participation.

First and most importantly, dynamics of *visibility* are fundamental to both artistic practices and surveillance – and a fortiori to surveillance art. As McKay (2013) points out, artists are trained specifically to look, to observe, to stare. Vito Acconci and Sophie Calle have pushed this skill to its limits in surveillance artworks in which they pry into the personal lives and effects of strangers. Other artists employ digital cameras and closed-circuit or wireless transmission networks, which make surveillance depersonalized, mediated and remote. For example, in McKay's video art made from filming strangers in public places, 'The closely zoomed images . . . suggest proximity, even uncomfortable over-proximity, and imply a level of intimacy with a stranger in a copresent space, when in fact, I was on another train some distance away' (McKay 2013: 344). This potential for distortion points to three 'models of visibility' – or ways of looking – identified by Brighenti (2010: 176). Visibility as recognition means that by looking at others, we acknowledge them as persons; ideally, it is reciprocal, with invisibility implying social exclusion. Visibility as control, in contrast, is non-reciprocal: here, unseen watchers look at others to gain power over them, often turning embodied subjects into abstract information; seeing becomes tracking. Finally, visibility as spectacle, also non-reciprocal, puts the viewer and the viewed on different planes, since 'the spectacle is a set of images detached from life and simultaneously falsely proposed as an illusory form of unity of life' (Brighenti 2010: 176). Here, the act of looking bridges yet separates the ordinary (watchers) from the extraordinary (watched).

While control might be the most common mode in surveillance as such, the other modes are often present in surveillance art, which implicates its viewers in patterns of recognition (evoking empathy for the watched) and spectacle (elevating quotidian images or data to a special, even sacred status). Much surveillance art plays with the different degrees of reciprocity between watcher and watched in each mode of visibility. For example, the Surveillance Camera Players, based in New York, devise short plays to perform in front of surveillance cameras, rendering their purpose absurd, in the manner of the court jester or fool, and turning the tables on the watchers who, if they notice the play, necessarily pass from the mode of control to viewers of a spectacle (Genosko 2006, Monahan 2006). Reciprocity raises questions about the ethics and legitimacy of surveillance art, which can be compelling precisely because it shows images that are usually not seen in public, thereby implicating its audience in an asymmetrical dynamic whereby they, as unseen watchers, also have power over the watched (McKay 2013).

The second idea of *location* refers to the ways that surveillance art fractures and compresses time and space. Surveillance extracts images or other data from one time and place and reproduces them at another, extending, expanding or displacing the field of vision and scope of knowledge of the surveillers. Crucially, it also disrupts the boundaries between public and private spaces or spheres. Indeed, since the tension between public and private is at the heart of debates about the expansion and regulation of surveillance, we could understand all surveillance art to be public art, if we define the latter as 'public' not by its siting, but by 'its engagement with the congested, cacophonous intersections of personal interests, collective values, social issues, political events, and wider cultural patterns that mark out our civic life' (Phillips 2004 [1989]: 192). Surveillance art puts the viewer in unexpected locations and dislocations.

Third, not surprisingly, *technology* is a crucial element of surveillance art. Cutting-edge surveillance technologies appeal especially to media artists, not only as tools they themselves can use to obtain desirable images or effects, but also as instruments – usually in the hands of the powerful – that can or should be hacked, dismantled, reassembled and subverted. Several artists use technological intervention to resist surveillance; for instance, artist-engineer Steve Mann makes wearable cameras that 'shoot back' at surveilled spaces (Genosko 2006, Monahan 2006). Artists' appropriations of surveillance technologies can open up promising cracks in systems that are usually regarded as monolithic, persistent and pervasive; they remind us to think about surveillance

> not as a merely technical process focused exclusively on recording images, tracking data flows and processing acquired and stored data, but also as a somewhat "artful" set of techniques, which may be more creative and, with Lévi-Strauss, more similar to *bricolage* than usually assumed.
>
> (Brighenti 2010: 175)

Surveillance as *bricolage* points to the fourth concept: *participation*. Albrechtslund and Lauritsen (2013) use Actor Network Theory to explain surveillance systems as networks that entail participation from heterogeneous actors, both human and non-human. Their starting point is that rather than being omnipotent and all-seeing, surveillance can instead monitor very small slices of life very thoroughly. This variability in scope raises the question of exactly how actors are 'enrolled' in surveillance networks: a GPS-enabled sports watch enrols participants into one sort of surveillance system, a CCTV camera into another. Therefore,

> participation must be studied specifically; how is participation established and maintained? What kind of resistance is present in the surveillance situation? Such a study might provoke criticism, as it is not possible to point out a priori powerful and powerless actors.
>
> (2013: 315)

We argue that art can draw attention in a compelling way to the participatory properties of surveillance systems and to the 'frictions and resistance' (Albrechtslund and Lauritsen 2013: 313) that they engender. By conjuring uncertainty out of everyday complacency, art can make plain the absurdity, the danger or simply the processes of existing surveillance networks. In this, surveillance art can be understood as constituting 'counter-surveillance', not in the strict sense of 'surveilling those who are doing the surveillance' (Marx 2003: 384) but in its broader definition as 'intentional, tactical uses, or disruptions of surveillance technologies to challenge institutional power asymmetries' (Monahan 2006: 516). We propose that *Time Transit* can be read in these terms as a work of public counter-surveillance art.

As contemporary commentary and the interviews we conducted show, *Time Transit* was both intended and understood to challenge surveillance (among other themes). Morgan found inspiration in two seemingly contradictory ideas about space (Morgan, Gelowitz and Benedicenti 2008): Foucault's (1977) image of Bentham's panopticon, the prison in which all inmates are watched at all times by unseen guards, and which, Foucault explains, they therefore internalize; and Lefebvre's 'illusion of transparency', the idea that our understanding of space as intelligible, 'as innocent, as free of traps or secret places' (Lefebvre 1991 [1974]: 28) actually masks the antagonistic history of its social production and significance. On the one hand, space can be engineered for total surveillance; on the other, it is riddled with invisible information. Modern surveillance technologies provide tools that span this tension: as they scan actual spaces, they send radio waves and data packets speeding through them, melding physical space with virtual space and what Dunne (2005) calls 'Hertzian' space. In *Time Transit*, the screens showing real-time and recent photos of the six bus stops were like windows into other parts of the city, while the website that tracked the bus along its route then embedded the whole moving installation into virtual space. In these ways, the artwork instated a visibility of control: people looking at the website could see the bus and the camera views, and people on the bus could see photos of people at the bus stops. However, since bus passengers were likely to see themselves or their peers in the photos, their way of looking also had an element of recognition; it was not entirely non-reciprocal. By playing with space, *Time Transit*, as its name suggests, also played with time, altering the location of the bus and its passengers. As one art critic put it:

> Collapsing present, future and past together, and folding one space – one spatial reality – within another, Morgan implodes the notion of the 'here' and the 'now' – and the notion of the city seen and understood only through a single physical position, a single point in time or a single social valence.
>
> (Anderson 2006: A12)

Thinking more broadly about its location, we should situate *Time Transit* in its historical moment. It was created during the new boom in electronic surveillance five years after 9/11, not long after the terrorist bomb attacks on public transport in Madrid

(11 March 2004) and London (7 July 2005). Transport companies were increasingly adopting surveillance technologies, most commonly CCTV and automatic vehicle locators to monitor schedules and employees and to identify 'suspicious' behaviour (Obermeyer 2007). Technologically, the telecommunications company SaskTel had only just rolled out its first wireless data network in 2005, and in fact was persuaded by Morgan and Gelowitz to waive the $5/MB transmission charge, since *Time Transit* would effectively test the capacity of the new infrastructure. Mobile phones had not yet spread to people from all walks of life (in Regina, at least) and were not yet 'smart'; texting was still relatively painstaking. Thus, although the installation's specifications sound banal today, at the time they required TRLabs engineers to develop 'bleeding-edge technology' (Gelowitz quoted in Beatty 2006: 23). For instance, it took an intern a month and a half to reverse engineer the proprietary software on a modem card to be able to pull the text messages off the phone networks and into the installation. As an 'early adopter' of GPS, wireless data transmission and mobile phones, *Time Transit* raised challenges to surveillance and dataveillance that have resonated well into the future.

Moreover, *Time Transit* enrolled participants such that they could express these challenges – and much more. For Morgan, the most significant feature of *Time Transit* was that it set up a platform for members of the public to interact with the art, to potentially use surveillance as material. Also with Gelowitz at TRLabs, Morgan had just created *Data s p a c e d* (2005), an audio installation in a passageway at the University of Regina, which captured the conversation of passersby then played it back to them a couple of hundred feet further along. In spite of initial complaints about the invasion of privacy, soon enough students were performing for the microphones and 'depositing' content – such as messages for friends – into the artwork. Morgan wanted *Time Transit* to open up a similar participatory space, yet as a public artist activating a public space, she did not control nor want to control the content of the work. So how did the public participate? Although people waiting at the bus stops might have put on a performance, the cameras' positions were not obvious and they were not zoomed in, so the visuals were a quieter element of participation. Instead, the focus of interaction was the artwork's capacity to receive and display text messages – some of which challenged the surveillance at play. About 206 messages sent to the installation, or 6.5 per cent of the total (excluding repeated 'spam' messages) questioned, praised or criticized the art project itself; for instance, 'I don't understand this', 'this time transit is idea is pretty kewl', 'What is the total cost of this art installation?'[2] Of these, 59 messages comment on its surveillant properties. Table 2.1 illustrates the range of reactions: some are trying to figure out the installation (#80, #3424, #3914); others are repulsed (#2615, #3656), politicized (#1862, #1053, #401), possibly jokey (#1053, #2500) or complicit (#65, #333 and #334, #755, #2706).

While the surveillant strategies of *Time Transit* did elicit some comment, by and large people expressed themselves on other matters. In the second part of our chapter, we analyse the corpus of text messages more thoroughly to show how public participation in *Time Transit* unsettled the usual atmosphere of the public bus and consequently changed how power circulated in this mobile public space.

Table 2.1 Selected messages about surveillance in *Time Transit*

#65	they should put a camera on the bus!!!
#80	I cann see you all as you enter this zone
#333	It's kind of like watching grass grow. . .
#334	But much more addicting ;)
#401	you've all become so comfortable being under surveillance
#755	i wish we could track every bus
#1053	Big Brother is watching you
#1862	just what we need - another advertising strategy. . .
#2500	This bus is being STALKED. Hoohaha.
#2615	this project is creepy. everyone is watching you
#2706	more cameras!
#3424	Someone is filtering these messages!
#3656	It sure feels good that we don't know who types these
#3914	the cameras arent realtime there fake

Civility and rudeness on Route 4

Like any type of space or situation, public transport spaces such as buses have their own particular social orders, regulated by recognizable norms of social interaction. Some are expressed in formal instructions to passengers on signs or audio messages, but most are implicit (don't sit next to a stranger when there are many empty double seats; reduce the space you take up as the vehicle becomes more crowded). More than simply taking care of the practical requirements of travel, these norms construct a kind of sociality among people who 'meet and move alongside for a short period of time', forming 'temporary congregations' (Jensen 2010: 400). As Wilson notes, 'the conditions of bus travel demand a highly attuned awareness of others, which is further shot through with particular obligations of travel' (2011: 639).

The physical conditions of bus travel include visibility – in the mode of recognition – and proximity: all bus passengers can see and are relatively close to each other, getting closer (but less visible) as the bus fills. Because of the proximity, recognition usually takes the form of 'civil inattention' (Goffman 1963: 84) – a glance that is focused enough to show that you acknowledge the other as a person, but fleeting enough to show that they (and therefore you) are not an object of unusual interest. Observing tacit norms and practising civil inattention on a bus go a long way toward maintaining an overarching atmosphere of civility. Broadly, civility means 'to have regard for the effects of our actions on others and to care for the space(s) we share with others, whether or not we are present in those spaces at the same time as others' (Fyfe, Bannister and Kearns 2006: 855). In the context of urban public spaces, it involves maintaining an attitude of reserve toward unknown people, which allows consideration for and communication with them while protecting everyone from the specifics and minutiae of social identities. It is a type of sociality that politely ignores or suspends social distance (difference) in order to cope with physical proximity (Simmel 1950: 409 ff.).

The work that maintains civility in public transport is often only made apparent when it is breached, when passengers contravene norms about seats or touching, when they press inappropriate attention on other passengers, when raucous, lewd or drunken behaviour erupts into the space.[3] These breaches may be repaired quickly and apologetically if they are accidental, or slowly and with more difficulty when there is misunderstanding or dispute over what is going wrong. Wilson (2011) describes several such encounters that unfolded during her ethnography of bus travel; what is striking is how rapidly and unpredictably they progress through different moments and assessments of discomfort, sometimes accumulating other breaches, before resolution. For instance, a racialized woman does not remove her bags from a spare seat in time to avoid an older white woman shaming her and then launching into a racist tirade; an unkempt man's loud appeals to other passengers are pointedly ignored until one of them realizes he has learning difficulties and helps him write a birthday card. During a bus journey, 'The space is continuously (re)negotiated and (re)ordered through mobile and constantly shifting practices of movement, behaviour, expectation, assumption, and assertion as people come and go. Differences are negotiated on the smallest of scales, individual itineraries come together – if only temporarily – and subjectivities are continuously (re)formed' (Wilson 2011: 646).

Time Transit clearly altered the social order of the Regina Transit bus on Route 4. Materially, it was a novel addition to the bus that must have been the focus of considerable attention and commentary. We never saw the installation, having heard about it only several years later, but we suppose that passengers' gazes were drawn to the screens (as we have witnessed in other transit spaces), and that it served as a 'triangulation point' (Lofland 1998: 39) enabling conversation between strangers, as do other shared third objects or events (children, dogs, buskers, delays). Socially, it provided an alternative forum for communication, which, crucially, circumvented the condition of visibility, since passengers could not see who had composed the messages that scrolled along *Time Transit*'s screens nor tell whether authors were sending them from the bus or the website. The installation enabled the public, momentary display of privately sent, anonymous messages, changing the conditions for social interaction on the bus.

We used qualitative data analysis methods to categorize the messages sent to *Time Transit* in terms of substantive content (what does the message express, describe or signify?) and communicative context (does it appear to have an intended individual or collective recipient? Does it stand alone, or is it a reply to another message or a turn in a longer dialogue?). The messages ranged from silly to serious, poetic to profound to profane. The stated aim of *Time Transit* was to 'encourage discussion about sustainable public transportation' (Figure 2.2). About a tenth of the messages did address public transit, but apart from a series of statements about Canada's and China's CO_2 emissions, they took a lighthearted tone, for instance, 'Route 4 and bus 600 is where the party is at!', 'This is like "Speed" for nerds'. And, repeatedly, the children's song 'The wheels on the bus go round and round' (which one day prompted the reply, 'Ohhh noooo! Not the wheels on the bus song!'). As mentioned above, some messages were comments on the project itself. Some referred to specific events (a Rolling Stones concert, a big football game, municipal elections) or places en route (University of Regina, SIAST technical college, McDonald's, Albert Street), illustrating how the installation was embedded in time and space. Another kind of message was the more or less facetious pearl of wisdom, such as 'Don't hate yourself in the morning, sleep till noon' or '"Love is not an emotion, it is your very existance"'. Pop-culture references – song lyrics, movie quotes – abounded. Reflecting the cutting-edge technology of the bus, many were memes from elsewhere in the digital world, including satirical factoids about American actor and martial-arts master Chuck Norris (a meme that apparently started in late 2005[4]), elaborate emoticons and nuggets of l33tspeak (from 'elite speak', a dialect incorporating numbers and symbols used by computer gamers and hackers). There are also more straightforward jokes. Most messages appear to address no one or everyone, including simple greetings ('Hi Bus People') sent perhaps just as tests. However, some are directed at individuals: 'Michelle.. you forgot your lunch!!!'; 'hey you with the red bag cheer up :):)'; others form conversations between named friends.

We turn now to the messages that, in our view, significantly affected the circulation of power in the bus: the rude ones. In this data subset, the ludic slides easily into the obscene. Many messages reek of toilet humour, and many others make more or less explicit or offensive reference to sex ('p3n1is' being a milder example). Since messages were displayed anonymously, such Rabelaisian bawdiness is not surprising. It seems likely that passengers were testing the limits of the system: just how rude could their messages be? The *Time Transit* team initially planned no limits on content except the 60-character maximum length; however, when bus company directors realized that the display could serve as a forum for insulting its drivers, they asked the project engineers to add a filter to block profanity. Of course, texters caught on to the censorship, so they put even more of their considerable creativity into finding ways to be rude without using the forbidden words ('HEY, HOOF HEARTED?', 'what da faaaaaaaaaaak?'). Filtering became a continual part of the project team's work: they ran regular manual checks to correct 'false positives', that is, remove texts that lacked forbidden words but were nonetheless offensive from the pool of messages for redisplay, and revised the first filter to add new words including deliberate misspellings and seemingly innocuous words that had secondary meanings (e.g. 'get off'). Because our dataset does not record at what point messages were blocked (the messages were time-stamped, but the filtering was not), it is hard to tell how many offensive messages were actually displayed. Still, by our analysis, 480 of the messages that were *sent* to the bus – or 12.2 per cent – can be classified as 'rude' in some way. Scattered between the pearls of wisdom and random acts of kindness, then, there were also random acts of unkindness.

Table 2.2 shows how these 480 messages break down. By far the largest category is 'sexualization' (n = 310), by which we mean messages that referred to sexual organs, sexual activities or sexual harassment (targeting specific individuals). 'Toilet humour' is the next largest category (n = 52), followed by insults (n = 37) that were not necessarily sexual or toilet-related but were directed at specific people or categories of people (mainly of the 'you smell' variety, plus insults of bus riders in general and the bus driver in particular). Homophobic messages (n = 28) were mostly along the lines of 'If you're wearing blue, you're gay :P'; since common slurs were blocked, we think only six messages using 'workarounds' such as 'ghey' and 'phag' remained after automatic filtering. Racist messages (*n* = 25) tended to be one-word slurs, which were blocked by the filter, or indirect comments such as writing in a racialized 'accent'. Because they contained no blocked words, the two messages that are arguably the most racist, because of their historical specificity – 'alright. new rule. only white people allowed on bus' followed immediately by 'The coloured section is at the back of the bus' – initially made it onto the screens, although they were later manually removed from the redisplay pool. A further 23 messages can be interpreted as rude because they refer to a stigmatized neighbourhood in Regina, North Central, which in 2006 had a significant Aboriginal population, a high incidence of poverty and dilapidated housing stock. Route 4 traverses North Central, so these messages were probably immediate comments on the bus's trajectory (e.g. 'You Are Now In The Hood ..Keep Clear Of The Windows' and 'You Are Now Out Of The Hood Congratulations You Made It'). Presumably to avoid reinforcing sociospatial stigmatization, the second (revised) filter added 'hood' to the list of blocked words. Finally, five messages are very violent, four of them being a threat against the bus driver (later manually removed).

Just classifying these messages is an interesting step (and a rare one, since interactive art installations are hardly ever systematically evaluated). But we want to take our analysis further. How can we explain these messages, and what might their effect have been on the social order of the bus, especially with respect to power dynamics?

Thinking through the conditions of production of the *Time Transit* messages helps explain their contents. Authors were guaranteed anonymity, so they could be as provocative as they liked without redress, yet they were also guaranteed an audience. The messages were ephemeral, scrolling away quickly, so inhibitions about leaving a permanent mark would not apply. They were short, facilitating sequential replies and cheeky banter between participants. Paradoxically, the

Table 2.2 'Rude' message analysis

Message	*n*	*% messages sent (n = 3960)*	*% 'rude' messages (n = 480)*
'Rude' messages	480	12.2	100.0
Blocked by first filter	*165*	*4.2*	*34.2*
Blocked by second filter	*100*	*2.5*	*20.7*
Sexualisation	310	7.8	64.2
Blocked by first filter	*105*	*2.7*	*21.8*
Blocked by second filter	*65*	*0.2*	*13.5*
Toilet humour	52	1.3	10.8
Insults	37	0.9	7.7
Homophobia	28	0.7	5.8
Racism	25	0.6	5.2
Stigmatised neighbourhood	23	0.6	4.8
Violence	5	0.1	1.0

filter itself may have encouraged rudeness, because once some authors realized it was there, they tried all the harder to circumvent it. Overall, the *Time Transit* messages were much less like ordinary text messages, which are usually sent between known interlocutors, than like a cross between toilet wall graffiti and certain Internet forums, such as message boards or the chat function within multiplayer video games, which facilitate anonymous, snappy and often very rude comments. These similarities fit the probable demographics of *Time Transit* contributors: young people and students, who were those most likely to be riding buses in car-focused Regina – especially Route 4, which passed two colleges. They were also most likely to own mobile phones, given their slow adoption in Canada, and more importantly, to be comfortable enough with technology to understand the point of the installation and send texts to it.[5] Regular contributors through the web-based interface were probably also young computer users, possibly video gamers. As with graffiti and anonymous Internet forums, being rude on *Time Transit* gave authors no penalty and, we imagine, considerable pleasure.

The parallels with other anonymous communication forums suggest not only how the rude messages were produced, but also what their effect was on the usual civility of the bus. Studies of (male) bathroom graffiti (Rodriguez and Clair 1999, Whiting and Koller 2007) and male-dominated online arenas such as sports message boards and video games (Fox and Tang 2014, Kian et al. 2011) underscore how they legitimate hegemonic masculinity. This is 'the form of masculinity in a given historical and society-wide setting that structures and legitimates hierarchical gender relations between men and women, between masculinity and femininity, and among men' (Messerschmidt 2012: 58), shoring up the overall oppressive relationship of men in general to women in general. Hegemonic masculinity depends on heterosexuality, heteronormativity and homophobia: in brief, the ideas 'that women exist as potential sexual objects for men while men are negated as sexual objects for men' (Donaldson 1993: 645). As per Gramsci's concept of hegemony, it never totally dominates, but rather, is applied and internalized unevenly and is always prone to resistance tactics. Nor is its power achieved through coercion, but 'largely through cultural ascendancy – discursive persuasion – encouraging all to consent to, coalesce around, and embody such unequal gender relations' (Messerschmidt 2012: 58).

The sexualized and homophobic messages show that *Time Transit* became, at times, a forum for the articulation of hegemonic masculinity. Messages espoused heteronormative male sexuality by referring to sexual organs and acts ('call me mr. seven incher', 'against the wall'), putting down homosexuality ('UR A phag!') and sexualizing women ('a girl's attractiveness is proportional to her availability', 'Double Jointed Girl, call me'). Some messages resembled street harassment ('Show me, YOUR CAMEL TOE!'). Non-sexual conversations could be abruptly hijacked by aggressive male sexuality:

> #675. The city needs more public dialogue.
> #677. The city needs more support for the arts.
> #678. this city needs more support for bukakke

Table 2.3 Messages resisting rudeness

#510	i hope no one posts anything mean
#642	Hey baby, show me your wits.
#1167	this city is unkind to minorities
#1403	NO SWEARING
#2266	Mmm, spam is tasty!
#3758	It Is Okay To Be Gay ^^

Indeed, *Time Transit* was sometimes entirely taken over by obscenity. By analysing content and turn-taking, we identified a conversation between three people calling each other Kevin, Vergil and Rue that lasted for four and a half hours one day. This conversation contains some of the rudest (and occasionally funniest) material in the dataset; it includes deliberately racist and sexual messages that were automatically blocked, as well as jokes, in-jokes and pop-culture references that are not necessarily uncivil – just opaque to other viewers, perhaps. The three apparent friends banter with each other; Rue is sexualized as female by Kevin and Vergil, who present as male. Perhaps these three rode the bus all afternoon; alternatively, they may have been contributing via computers, which would explain the speed of replies and the length of the conversation. (No markers in the dataset differentiate between mobile phone and website messages except obvious content, such as 'Hey, baby, I am watching this from Tokyo, congratulations!') Very few other types of messages got through that afternoon, and the three returned to take over again later during the installation, though for shorter periods. We hypothesize that Kevin, Vergil and Rue were already heavy computer or cellphone users, ready to appropriate digital arenas for their own rowdy, bawdy enjoyment.

Given that *Time Transit*, like bathroom stalls, offered the space to reply, it seems surprising that there was so little resistance to incivility. We discern only six messages (Table 2.3) that seem to raise objections, four direct (#510, #1167, #1403, #3758) and two wittier (#642, #2266). (#3758 would have been automatically blocked, because it contains 'Gay', but was later manually replaced in the pool.) Perhaps messages went by so quickly that those who might have responded preferred instead to ignore them. Perhaps no significant culture of quashing online misogyny, racism and homophobia had yet built up in small-town Canada, as it arguably might have done at the time in bigger cities or more widely today, when controversies of sexism and racism in digital and high-tech environments make breaking news.

Still, we argue that the rude messages in *Time Transit* had problematic consequences for power dynamics in the public space of the bus, especially when they dominated the content. First, they likely made some passengers feel personally uncomfortable. Unlike bathroom-stall graffiti or Internet chatrooms, where messages are viewed in a sequential, solitary way, by one person at a time, the principal audience here was collective and co-present, sharing the space of the bus.[6] De Leeuw notes:

> The spaces of public transit, be they station hubs, the buses and trains themselves, the routes and tracks and lines upon which travel the various vehicles, or the seats for which people jostle, are also very much alive spaces, full of human emotion, relationship, and expectation. As Attoh observes, they are places 'packed with people' … Each of these people, purely by virtue of sharing the places and times of public transit, is engaged in a relationship with each other.
>
> (de Leeuw in Attoh et al. 2011: 290)

The passengers on Route 4 read the *Time Transit* messages each with their own emotions and expectations, and their reactions must have varied. The exact same texts might have been entertaining for some, but tiresome, puzzling, embarrassing or worrying to others. Moreover, the *interrelationship* of passengers would have been disturbed by some messages. We suppose that some women would have felt threatened by seeing themselves objectified in real time by unidentifiable, presumably male, riders; likewise, homophobic and racist messages probably made the space feel less safe for LGBTQ+ people and racialized minorities.

Second, the uncivil messages that populated and at times dominated *Time Transit* underscored the symbolic exclusion of certain groups from public space. Along with scholars who have analysed toilet graffiti (Bruner and Kelso 1980, Rodriguez and Clair 1999, Whiting and Koller 2007), we maintain that even though anonymous open forums seem to encourage 'say[ing] whatever, however, and whenever, to whomever' (Rodriguez and Clair 1999: 2), they must not be read as scattershot explosions of passion, but rather as communicative discursive spaces that reflect and reinforce their local context and broader societal structures. Thus, the misogynistic and homophobic messages in *Time Transit* 're-instantiate hegemonic masculinity as constructed in a hierarchical relation to femininity' (Whiting and Koller 2007: 27). The racist messages, though far fewer in number, would have had the same powerful effect of 'disciplining otherness' (Rodriguez and Clair 1999: 12), using hostility to reassert white dominance of public space. Moreover, the lack of contestation of incivility had the communicative effect of giving tacit approval to the dominant discourse (Kian et al. 2011).

This means that whether or not any threat was personally felt as 'real' or likely to be acted on, the rude messages disrupted the collective social order of public transit by drawing attention to differences – of gender, sexuality and racialization – that mark inequalities of power between different social groups. Rather than suspending difference and acting as if all were equal, as per urban civility (Simmel 1950), the rude messages drew explicit social boundaries. They went far beyond the normal bounds of communication in public space, constituting 'uncivil attention' as opposed to civil inattention. By providing an anonymous and ephemeral forum for self-expression, *Time Transit* allowed participants to breach the principles of mutual visibility and recognition that usually regulate public transport spaces. It therefore had the (unanticipated) effect of enabling the articulation of unequal power relations that marginalize certain social groups in public space.

Conclusion: incivility in public art

In this chapter, we have explored how the power dynamics at work in Kim Morgan's *Time Transit* installation (Regina, 2006) unsettled the social order of an everyday urban space. First, we argued that *Time Transit* can be read as a work of public counter-surveillance art. By appropriating and repurposing cutting-edge surveillance technologies of cameras, GPS and dataveillance in an ordinary public bus, *Time Transit* critiqued the non-reciprocal visibility of control that surveillance systems usually instate. By putting surveillance photographs on public display in the bus, it introduced elements of recognition and spectacle into this dynamic of visibility. It created a platform for participation, as viewers could send text messages to the installation screens. The artwork refracted and multiplied the locations of the bus, allowing passengers to view and reference other places or moments. It stitched together and pulled apart time and space, and public and private spheres of interaction. Our application of sociological concepts thus helps unpack the social context of a cutting-edge work of public art, especially the macro-level power dynamics in which it was embedded. Second, we interpreted the interactive content of *Time Transit* as facilitating breaches of urban civility. The anonymous, ephemeral text messages sent to *Time Transit* resembled not so much a serious public dialogue as a carnivalesque explosion of mobile graffiti, by turns entertaining, obscene or offensive. We suggest that when used to express sexist, racist, homophobic, violent or otherwise rude messages, the interactive element of *Time Transit* had the effect of reinscribing inequalities between privileged and subaltern social groups. A public artwork that appropriated surveillance technology in order to interrogate power relations in society at large was itself appropriated by members of the public in ways that troubled power relations at the micro-level.

Our urban anthropological approach involves analysing public art at several levels: in terms of broader societal trends contextualizing the art (burgeoning surveillance), particular social dynamics encapsulated within the art (visibility, location, participation, technology) and evidence of micro-level social encounters between art and the public (the text messages). Our analysis raises critical questions about the situational impact of public art and provides a basis for the interpretive answers we offer in conclusion. It is important not to exaggerate the effect of incivility in *Time Transit.* Messages were fleeting and the audience, fluctuating (though probably also regular); civility may have been restored with the very next message. However, we should still ask whether the rudeness was worth it. Put bluntly, should public art be a venue for the expression of sexism, racism, homophobia and violence – however small a proportion such discourse is of the overall interactivity? Should it propagate poop jokes? If we cautiously answer 'no', we may end up with harmless, mediocre public art made to a 'minimum basic standard' (Phillips 2004 [1989]), always trailing behind the artistic avant-garde. We think that public art indeed should risk provoking controversy and causing offence as it strives to engage with critical issues.

In the case of *Time Transit*, it is easy to argue for its macro-level value as a work of counter-surveillance art: deconstructing ubiquitous surveillance seems never more urgent. But the artwork is also valuable for its micro-level effects on social relations. First, there is value in providing an interactive platform, however it was used or abused; the playful, subversive messages reveal the creativity usually buried in everyday urban spaces. Second, it can be salutary to expose societal fault-lines that are normally concealed, to remind us of who dares speak and who stays silent when the strange liminal space of a public yet anonymous forum opens up. Kim Morgan's *Time Transit* thus realized Phillips's (2004 [1989]: 192) ideal of public art: that it should 'not find its meaning through its situation in a forum, but . . . create the forum for the poignant and potent dialogue between public ideals and private impulses, between obligation and desire, between being of a community and solitude.' Wonderfully, and rather by accident, its creators also retained and granted us access to the data that enabled us to ground our social analysis in empirical evidence. Similar collaborations between artists and social scientists can make for highly productive analyses of the power dynamics that circulate in public art encounters.

Acknowledgements

The research for this chapter was funded by an award for Research-Creation in the Fine Arts (#848-2010-0019) from the Social Sciences and Humanities Research Council of Canada. We thank Kim Morgan and Craig Gelowitz for generously sharing their data and ideas about *Time Transit*, and Nethra Samarawickrema and Emily Fraser for providing excellent supplementary research assistance.

Notes

1 The other *Time Transit* team members were programmer Bill Friedrich, media artist and research associate Lee Henderson, graphic designer Jane Uttaranakorn and engineer Laura Wiley.
2 All messages are reproduced exactly as written.
3 One reason that surveillance systems such as CCTV are installed in buses is, of course, to prevent or at least record any such breaches, including conflicts between passengers or with drivers (Obermeyer 2007).
4 http://en.wikipedia.org/wiki/Chuck_norris_facts, accessed 1 September 2015.
5 In 2006, Canada had only 57 mobile phone subscriptions per 100 people, compared to 76 in the USA, 95 in Australia, 116 in the UK and 136 in Italy (Source: World Bank World Development Indicators via Google's Public Data Explorer, www.google.ca/publicdata/directory?hl=en, accessed 1 September 2015). If *Time Transit* existed today, we suspect that a wider range of people would engage with it. While they might be more articulate, they might also be more circumspect about their contributions, due to the cross-tabulation possibilities of 'big data'.
6 We suppose that the audience for the website, which showed the currently scrolling message, was small.

References

Albrechtslund, A. and Dubbeld, L. 2005. The plays and arts of surveillance: studying surveillance as entertainment. *Surveillance & Society*, 3(2/3), 216–221.

Albrechtslund, A. and Lauritsen, P. 2013. Spaces of everyday surveillance: unfolding an analytical concept of participation. *Geoforum*, 49, 310–316.

Anderson, J. 2006. Artist remaps city in new ways. *Leader-Post*, 29 November, pp. A12–A13.

Attoh, K., Butz, D., Hones, S. and De Leeuw, S. 2011. The Bus Hub. *ACME: An International E-Journal for Critical Geographies*, 10(2), 278–292.

Barnard-Wills, K. and Barnard-Wills, D. 2012. Invisible surveillance in visual art. *Surveillance & Society*, 10(3/4), 204–214.

Barone, T. and Eisner, E. 2012. *Arts Based Research*. Thousand Oaks, CA: SAGE.

Beatty, G. 2006. Kim Morgan: artist's high-tech transit jam merges intermedia and engineering. *Prairie Dog*, 26 October, p. 23.

Bennett, C., Lyon, D., Haggerty, K. and Steeves, V. (eds) 2014. *Transparent Lives: Surveillance in Canada*. Edmonton: Athabasca University Press.

Brighenti, A. 2010. Artveillance: at the crossroads of art and surveillance. *Surveillance & Society*, 7(2), 175–186.

Bruner, E. and Kelso, J. 1980. Gender differences in graffiti: a semiotic perspective. *Women's Studies International Quarterly*, 3(2–3), 239–252.

Donaldson, M. 1993. What is hegemonic masculinity? *Theory and Society*, 22(5), 643–657.

Dunne, A. 2005. *Hertzian Tales: Electronic Products, Aesthetic Experience, and Critical Design*. Cambridge, MA: MIT Press.

Foucault, M. 1977. *Discipline and Punish: The Birth of the Prison*. New York: Pantheon Books.

Fox, J. and Tang, W. 2014. Sexism in online video games: the role of conformity to masculine norms and social dominance orientation. *Computers in Human Behavior*, 33, 314–320.

Fyfe, N., Bannister, J. and Kearns, A. 2006. (In)civility and the City. *Urban Studies*, 43(5/6), 853–861.

Genosko, G. 2006. (Im)Possible Exchanges: the arts of counter-surveillance, in *Canadian Cultural Poesis: Essays on Canadian Culture*, edited by G. Sherbert, A. Gérin and S. Petty. Waterloo, ON: Wilfrid Laurier University Press, pp. 31–50.

Goffman, E. 1963. *Behaviour in Public Places: Notes on the Social Organization of Gatherings*. New York: Free Press.

Jensen, O. 2010. Negotiation in motion: unpacking a geography of mobility. *Space and Culture*, 13(4), 389–402.

Kian, E., Clavio, G., Vincent, J. and Shaw, S. 2011. Homophobic and sexist yet uncontested: examining football fan postings on Internet message boards. *Journal of Homosexuality*, 58(5), 680–699.

Lefebvre, H. 1991 [1974]. *The Production of Space*. Malden, MA: Blackwell.

Lofland, L. 1998. *The Public Realm: Exploring the City's Quintessential Social Territory*. Hawthorne, NY: Aldine de Gruyter.

Lyon, D. 2003. *Surveillance after September 11*. Malden, MA: Polity Press/Blackwell.

McKay, C. 2013. Covert: the artist as voyeur. *Surveillance and Society*, 11(3/4), 334–353.

Marx, G. 2003. A tack in the shoe: neutralizing and resisting the new surveillance. *Journal of Social Issues*, 59(2), 369–390.

Messerschmidt, J. 2012. Engendering gendered knowledge: assessing the academic appropriation of hegemonic masculinity. *Men and Masculinities*, 15(1), 56–76.

Monahan, T. 2006. Counter-surveillance as political intervention? *Social Semiotics*, 16(4), 515–534.

Morgan, K., Gelowitz, C. and Benedicenti, L. 2008. The public space as an interface for technology research and art: a study and implementation of two interdisciplinary collaborations between engineers and artists. *The International Journal of Technology, Knowledge and Society*, 4(3), 65–72.

Obermeyer, N. 2007. Moving violations: data privacy in public transit. *Geographical Review*, 97(3), 351–364.

Phillips, P. 2004 [1989]. Out of order: the public art machine, in *The City Cultures Reader* (2nd edn), edited by M. Miles, T. Hall and I. Borden. London and New York: Routledge, pp. 190–196.

Rodriguez, A. and Clair, R. 1999. Graffiti as communication: exploring the discursive tensions of anonymous texts. *Southern Communication Journal*, 65(1), 1–15.

Simmel, G. 1950. *The Sociology of Georg Simmel.* New York: Free Press.

TRLabs. 2006. Time Transit: Regina Transit to Host TRLabs Mobile Art Installation. *TRLabs 2006 Review.* Regina: TRLabs, 22–23.

Whiting, S. and Koller, V. 2007. Dialogues in solitude: the discursive structures and social functions of male toilet graffiti. Centre for Language in Social Life (CLSL), Lancaster University: CLSL working paper series, no. 126.

Wilson, H. 2011. Passing propinquities in the multicultural city: the everyday encounters of bus passengering. *Environment and Planning A*, 43(3), 634–649.

3 'Awaken the Dragon'

Participatory art-making and the grassroots in authoritarian Singapore

Jason Luger

Introduction

> 'Awakening the Dragon' is, I guess, a consciousness that – Singapore is developing so fast – we should really stop and appreciate things while they are here, or you know, before something happens to it. Maybe – take ownership of that place.
>
> And I guess that is what the 'Dragon Kiln' is about. We wanted people to take ownership of the kiln so they feel like – this is their heritage, and they should protect it. Or they should at least – know the story.
>
> (Interview, Festival Curator, February 2014)

In 2015, two remarkable events happened in Singapore: in March, the founding Prime Minister, Lee Kuan Yew, often credited with building the groundwork for the successful city-state, died. Then, in August, Singapore celebrated its 50th birthday as an independent nation, triumphantly reflecting upon its unprecedented rise from a small island without natural resources to a leading global city. Yet despite these moments of national unity, undercurrents of tension and discord continue to bubble beneath the surface; what Lefebvre (1991[1974]) called the 'seething forces' of dissent make themselves known through periodic ruptures and eruptions. In 2011 a contested general election saw the ruling authoritarian government face its first real challenge in a generation; in 2012, a labour strike made international news and forced Singaporeans to reconsider their reliance on imported manual labour; and in 2013, a race riot broke out in 'Little India' when a Chinese bus driver hit and killed an Indian pedestrian. All of these ruptures are unprecedented and exposed a more cacophonous side of 'harmonious' Singapore, and starkly highlight the city-state's current socio-cultural turning point.

An enlivened grassroots is one vehicle through which critical questions emerge and are negotiated; the arts and cultural realm is a theatre within which the grassroots can operate particularly visibly within authoritarian confines. Traditional protest and dissent in the form of street marches or rallies is generally restricted (confined to certain areas, and must be pre-approved by authorities, remaining within pre-defined 'out of bounds' markers; see Luger 2015, 2016). Therefore, art-making – and the artistic encounter – is one way that critical voices can emerge through the nooks and crannies of the authoritarian structure. This art-generated potential is certainly not unique to Singapore, but particularly highlighted due to the lack of other such critical spaces and sites of civil society rupture. The negotiations between the artistic grassroots, the authoritarian state and the built environment are a complex performance, characterised by bilateral subversions and appropriations of what the artistic encounter can – and should – look like in such a context. The example of the 'Awaken the Dragon' festival represents one such complex, contradictory encounter: a state-generated and state-funded arts festival involving participatory ceramics-making yet involving many critical voices and, ultimately, a critical challenge to the state's urban-development plan. This chapter thus interrogates the questions that emerge of (a) *whether critical artistic responses to state policy are indeed a deliberate result of such policy*, or if (b) *these critical responses emerge largely against – and independently from – state policy architecture*. Indeed, the relationship between deliberate and non-deliberate activism, the simple act of making art, and the state-aim of fostering a more participatory artistic grassroots – may not be so straightforward; intentions and impacts may be left open, uncharted, and spontaneous. Therefore, this case study adds depth to the ongoing – yet still emergent – conversations about the state, art, activism and urban space.

This chapter explores the networks involved in the landscape of the artistic grassroots, state arts policy nexus, the Singaporean 'heartland' (see Tan 2008) and how these encounters are played out upon/within the urban environment through the lens of the 'Awaken the Dragon' festival. The 'Awaken' festival demonstrates how a state-sanctioned participatory-public art event, led by grassroots artists (some of whom are critical of state policy) can generate particular encounters and outcomes that are capable of reshaping policy, produce new networks and alliances, and re-appropriate and reclaim the urban environment and cultural heritage. Furthermore, this chapter critically engages with the literature on 'cultural activism' (see Buser et al. 2013) by questioning the conception of strategic networks and alliances as portrayed in many Western contexts. In Singapore's unique quasi-authoritarian setting and city-state scale, space, activism and public art all take on different forms and require different understandings. Atypical contexts such as Singapore, with its hybrid and peculiar power structure and island scale invite revisiting, and challenging, the literature on the 'renewed' left and the new geographies of resistance and activism (Amin and Thrift 2013, Davidson 2013, Harvey 2012, Mayer 2009). In particular, this chapter engages with new conceptions of deliberate versus non-deliberate activism, the notion of activist-autonomy, and the ways in which the activist performs criticality simply through everyday activity (Chatterton and Pickerill 2010, Katz 2004). Finally, this chapter frames the nooks and crannies of the critical art encounter enabled by the 'awaken' festival within Foucault's (1982) conception of 'governable'/'ungovernable' space and within Lefebvre's (1991[1974]) 'spatial trilogy' in which space is represented (as an urban plan); is representational (through practice); and is lived and practised both through and due to the artistic imagination.

First, this chapter will briefly outline Singapore's 'cultural turn' (Luger 2015) and the context of the artistic grassroots in the city-state. Singapore's rapid transition from 'third world to first' (Lee 2000) has seen a recent emphasis on arts and culture alongside more traditional state investments such as infrastructure and education. This overview includes a contextualisation of the blurry line between activism, the state and artistic space within Singapore's quasi-authoritarian context.

Following this contextual overview, this chapter engages with the 'Awaken' festival, which illustrates how state (top-down) arts and culture policy becomes performed, and how the performance of such policy both meets policy aims and challenges them: did the state intend to save the dragon kiln when it agreed to fund the 'Awaken' festival? Therefore, art policy becomes a 'Pandora's Box' when spontaneous grassroots occurrences and encounters move the production of [artistic] space in new and unplanned directions.

These observations stem from a larger research project conducted between 2011–2015 which explored the critical arts and civil society in Singapore, involving semi-structured interviews (N = 30) with artists, activists and state officials; participant/participatory observation, and policy/discourse analysis. The selected empirical examples are based on embedded field research conducted in Singapore during 2012–2013, with a follow-up visit in 2014. Due to political sensitivity, names are not used and in some cases pseudonyms are used as replacements.

Contextualising Singapore's (cultural) turning point

Singapore is an island-nation city-state located in the Malay-Indo Archipelago, adjacent to the busiest shipping channel in the world. This location has been Singapore's strategic advantage since its founding by the British in 1819. It gained independence in 1963, when it was briefly joined with Malaysia, but has been an independent nation since 1965. Singapore has transitioned from 'third world to first' (in the words of its founding father, the late Lee Kuan Yew) in a single generation, and is now among the world's wealthiest nations (Economist Intelligence Unit, 2015) as it has become a leading global financial and business services centre. The city-state has a small (but growing) expatriate community of Westerners, but is majority ethnically Chinese (more than 75 per cent). These Chinese are both recent arrivals from Mainland China and also descendants of the 'Straits Chinese' who have settled in the region for hundreds of years. There are substantial Malay (11 to 12 per cent) and Indian (8 to 9 per cent) minorities, who (unlike the Chinese), are primarily Muslim, Hindu and Sikh. This tapestry of colonial and post-colonial history and culture, cosmopolitan and international ethnic and religious communities gives Singapore a rich, complicated and sometimes tense potpourri of identities and a fluid conception of 'nationhood'. This conflicted self-identity – a majority Chinese, English-speaking nation within the broader Malay (Muslim) world, with a global outlook – is one of the government's rationales for Singapore's continued single-party authoritarian rule, led by the People's Action Party (PAP). The PAP has perpetuated an approach to nation-building that values economic growth, efficiency, harmony, stability, meritocracy and a strong rule of law that is sometimes seen as harsh by external observers (see Chua 1997, 2011). One of the results of the government's high level of control is a heavily top-down policy and planning system, in which more than 85 per cent of land is state-owned and development is carried out by state-owned corporations (Shatkin 2014). This applies also to the crafting of urban space, as development has resulted in the loss of many historic structures and neighbourhoods as Singapore has grown larger (and wealthier). Some historical districts (such as Chinatown) have been preserved which reinforce national identity or, as in the colonial-era Raffles Hotel, upscale tourism (see Luger 2016). Art and cultural initiatives are likewise highly planned, crafted and directed downward by various state agencies and departments, subject to a labyrinthine maze of checks and balances, censorship reviews and approvals.

Paralleling Singapore's rise into a first-tier city has been the government's embrace of a cultural turn (Chang 2008, 2015, Chang and Huang 2009, Chang 2015, Kong 2012, Luger 2015, 2016, Ooi 2008), particularly its aim to become a 'Global City for the Arts' (National Arts Council 2012). The 'arts' are envisioned in state policy documents as an economic tool to appease, attract and retain cosmopolitan residents and visitors to Singapore – but also as a nation-building tool, in which Singapore's various communities (the 'heartland' and the 'cosmopolitans', as bifurcated by KP Tan 2008) can come together around a shared identity. Some examples of the *economic* cultural investments include massive new museums, entertainment destinations and a push for more tourism expenditure (from the Marina Bay attractions to the relaxing of the casino ban in 2005). Examples of the nation-building cultural investments include the preservation of ethnic neighbourhoods such as 'Chinatown', 'Little India' and 'Kampong Glam' (which has a Malay identity); the annual 'National Day Song', which is composed uniquely each year for Singapore's birthday; and island-wide arts festivals that take place in the (public) housing estates where most Singaporeans live and around issues/areas of common history and heritage. This last aspect of the 'cultural turn' was encapsulated in 2012's 'Arts and Cultural Strategic Review', in which art and culture were to be for 'everyone, everywhere, everyday' (National Arts Council 2012: 26).

Within Singapore's cultural turn, artists and arts spaces have been given a prominent, if conflicted, role to play. Artists have been provided housing and performance space, and funding streams have been directed toward art groups in a variety of ways. Small theatres such as 'The Substation' and smaller theatre groups such as 'Wild Rice' receive state subsidies; other arts groups remain independently funded but will sometimes engage with state funding for specific projects or initiatives (such as the 'Awaken the Dragon' festival, spearheaded by a group of independent artists).

Reading – and reinterpreting artistic criticality in Singapore's context

At the same time that the arts receive space, funding and (relative) freedom of expression, government entities such as the Media Development Authority

(MDA) and the Censorship Review Committee (CRC) retain restrictions on content and creative content. Plays, visual art or other displays that violate Singapore's 'out of bounds' markers – a grey area, but generally forbidding content that deals controversially with racial, religious, ethnic or political themes – are subject to being banned and de-funded (see Ooi 2008, 2009, 2010, Yue 2007) or even face prosecution, evidenced by the plight of several critical bloggers in 2014–2015. This creates an environment that is – contradictorily – highly supportive of the grassroots arts ecosystem by providing space and funding, but also highly restrictive, since art must stay within certain (and constantly changing) confines. Thus, civil society plays a complex, pragmatic and nuanced game in terms of pushing boundaries and challenging the power structure, without pushing too far. This requires a careful reading of grassroots activism in Singapore and a reconceptualisation within the city-state's unique context of what the geographies of – and performance of – critical art-making looks like.

The arts in Singapore can broadly be divided into two branches: the 'cosmopolitan' (elite) arts, and the 'heartland' (grassroots and/or non-elite) arts. Sometimes, these branches touch each other: the grassroots, generally highly educated with rich social capital – is often involved in artistic cosmopolitanisms. Art spaces, places and events also take on this bifurcated nature. Elite events such as the Singapore Biennale, or places such as private for-profit galleries, cater to the global art market and those wealthy enough to purchase/consume art. Tickets to the Broadway and West End shows that come to (the theatre and casino complex) Marina Bay regularly cost hundreds of dollars. 'Heartland' arts, on the other hand, include participatory festivals such as the 'Void Deck Festival' which takes place in the HDB (state-housing) estates of working and middle-class Singapore and include events such as group dances, morning tai chi, sing-alongs, and ethnic-themed arts and crafts. My field research interviews and observations indicated that these two branches of the arts often exist in separate planes: those who participate in finger painting at their housing estate do not often attend exclusive art-gallery openings; those who attend the Biennale probably do not attend the 'Void Deck Festival'. The division of the arts and within the art-scape is emblematic of the wider divides and fissures in Singaporean society which increasingly challenge the status quo. This comes as the (current) prime minister publicly stated that 'we have no internal divides' (observed at Chatham House, on the record, March 2014).

Meanwhile, there have been undercurrents of dissent and discord. Singapore's high cost of living has made housing a contentious issue. Inequality has risen as wages have stagnated for many Singaporeans as highly paid expatriates have taken top jobs, and lower-paid economic migrants have taken manual labour positions: Singapore has no minimum wage. Anti-foreign sentiment has risen, causing some to question the sustainability of the economic approach that is highly dependent on imported labour. There have been, along with this, racial and ethnic tensions not observed in many years: the English-speaking Chinese majority and the Malay-speaking minority have been challenged by new arrivals from China who speak mainly Mandarin Chinese. 'Singlish', a local English variant which incorporates regional elements, has been discouraged by education policy-makers in favour of the business languages of 'Queen's English' and Mandarin. Meanwhile, race riots erupted for the first time since the 1960s in 'Little India' in December 2013 when a Chinese bus driver hit and killed an Indian pedestrian who had been trying to board the moving bus. This occurred on a Sunday, which is a day off for tens of thousands of migrant workers who congregate in 'Little India' to socialise. The riots symbolise the cracks that have appeared in Singapore's tightly controlled, stable socio-cultural fabric. The riots are also emblematic of the battles over space that define the tension in Singapore – this space includes urban space (in 'Little India', migrants jostle with locals and tourists for use and occupation of the same square mile), but also ideological space (the Singaporean blogosphere is loud, crowded and often antagonistic, with cyber-citizenry free to engage in ways not often possible on the street).

Human rights are another issue of concern for Singaporean activists: continued censorship of artistic content deemed inappropriate by authorities has generated a loud debate within civil society. Singapore's retention of several (British) colonial-era laws – ranging from the illegality of sodomy (Penal Code 377a) to the state's ability to detain without cause – generate loud debates (in coffee shops, on the street and in cyberspace).

A final issue of contention, and highly relevant to this chapter, is the use and development of urban space on a land-starved island. A recent 'Population White Paper' (Singapore Government 2013) outlined the need to plan for 7 million people by 2030, requiring the construction of several large 'new towns'. These plans have generated controversy as many green spaces are threatened by development, including 'Bukit Brown Cemetery', the largest ethnically Chinese cemetery outside of China and an important historic and cultural site for the Chinese diaspora and Singapore's Chinese majority. Another historic space that is threatened by development is the site of the Jalan Bahar Clay Studios, the site of Singapore's last-operating dragon kiln. Dragon kilns historically were important features of Singapore's ceramics industry, which has moved from (what was) a major export to the domain of a select group of hobbyists. It is the state's plans for destruction of the dragon kiln that spurred the artist-activist movement that became the 'Awaken the Dragon Festival', in 2012/2013.

The critical (artistic) grassroots in Singapore

Artists have been leading forces within the activist grassroots in Singapore, for a number of reasons: first, the arts are given (relatively) more space for critical views than traditional methods. Street protests and political rallies are forbidden, outside of those that are pre-approved and take place in 'Speakers' Corner', a central square. Second, artists are able to tap into funding – an externality of the state's 'cultural turn' discussed previously. Third, Singapore's small geography and tightly connected activist networks mean that artists are highly networked with each other, and also highly connected to various activist groups (around a range of issues and themes) that may not be the case in other (larger or more spread-out) places. Artists are therefore present and active in many (if not *most*) local activist causes, and the arts and activism join together for both strategic and circumstantial reasons: arts, culture and activism are blurred in a constant performance that includes mutual co-option; strategies to survive and practical tactics maintain resiliency by staying within authoritarian boundaries (see Lim 2012, Rae 2011).

The city-state has a long tradition with 'forum theatre', a method of participatory political art with leftist origins. In the 1980s, small-scale theatre groups such as 'The Necessary Stage' often used 'forum theatre' to explore a variety of themes which crossed some of Singapore's most controversial boundaries, such as race, ethnicity, religion and sexuality. In 1989, the government cracked down on 'forum theatre', and the art form was banned for several years. However, recognising 'forum theatre's' ability to constructively involve public participation in national ideology and the performance of nation-building, 'forum theatre' has been re-introduced and appropriated by the state for state-directed (rather than grassroots-led) uses (Tan 2013, on this strange reinstating and co-optation). Some of the same theatre companies once banned and facing serious penalties for using 'forum theatre' are now employed by the state to use 'forum theatre'. This very Singaporean paradox clearly demonstrates the careful reading necessary of artistic criticality and its agents and networks in Singapore: 'forum theatre' contains distinctly radical potential, and has its origins in Marxist movements in South America – yet the way it has been co-opted, used and deployed by the authoritarian state means that it now falls under the state umbrella as a tool for nation-building and social harmony (rather than dissent or social activism).

Grassroots arts groups have been central to activist issues over the past generation in Singapore, ranging from censorship of the theatre to civil and human rights. The annual 'Pink Dot' LGBTQ gathering at Hong Lim Park always features artists as keynote speakers (such as Samantha Lo, the openly gay 'Sticker Lady' graffiti artist). Additionally, art displays have often taken on other aspects of Singapore's political-socio-cultural milieu; 'Sticker Lady', the street artist, left a trail of playfully subversive stickers in public places around Singapore in 2012–2013 that critiqued topics such as Singaporeans' materiality, lack of appreciation for history and bruising work ethic. She was jailed, but later released.

One arts group, called 'Art-Post', is often at the forefront of cultural-activist causes. 'Art-Post' operates within Singapore as a sort of experimental gallery without walls; they organise a variety of different events around the island, some with decided activist messages, others without. That 'Art Post' does not have a permanent home, and exists in a fleeting and quickly moving manner, reflects their constant state of precariousness as a small-scale, grassroots arts group that receives very little state support. At the same time, 'Art Post' has, over their existence, remained resilient and found various ways to spread their message and spread critical art. 'Art Post' frequently organises the 'Really Really Free Market', an event held in various locations around Singapore, influenced by other global collectivist movements. 'Art Post' was also a principal organiser of the 'Awaken the Dragon' festival, helping to curate and to lead the clay-making sessions around the city-state. For 'Art Post' founders, space itself forms the basis of their work, aims and approach, rather than any particular cause or social-political issue. The interrogation of and creative use of space, and determining new and different spatial possibilities, is not without meaning in space-constricted Singapore – particularly given the high degree of central planning and state ownership of space:

> The thing with the Dragon Kiln – I think it was kind of, in a way – unintended from the start? Because it – so happened that the [Government] is going to build this highway, and then we thought that we should do something about it … I guess we are looking at spaces which had various possibilities for us. Yea. So I think – for every space, wherever it is, there is always this kind of possibility and potential – and also, the social interactions and the, you know the life of the city, life of the people there –there are always things which are interesting.
>
> (Interview, 'Art Post' founder, December 2012)

The 'Awaken the Dragon' festival, financed partially by the National Arts Council and led by a group of Singapore-based artists, is one compelling example that represents the complex and somewhat contradictory environment of the *public art encounter* in Singapore. Festival participants are simultaneously reifying the state by taking part in a nation-building performance, which involves connecting with a shared regional heritage of kiln-fired ceramics; yet, they are also working together to re-claim a piece of Singapore's urban landscape that is threatened by state-led development plans. The success of the festival in generating much national and international publicity, combined with its rare ability to connect with a wide spectrum of Singaporeans (one of the aims of state arts policy), has meant that the state has had to change course and spare the historic dragon kiln from demolition in the name of development. This may not be a straightforward grassroots 'victory', but such changes of state planning and development policy are very rare in Singapore, where economic development and political stability usually come before sentimental attachments to historic urban or culture space, and backtracking on policy is often interpreted as an embarrassing loss of face for the meritocratic policy elite.

'Awakening the Dragon': making clay, shaping identity – and challenging policy

The Jalan Bahar clay studios are located on the fringe of Singapore, some ten miles away from the City-Centre, adjacent to the campus of the Nanyang Technological University. At the clay studios, in an open-air space with a ceiling canopy, is the last-operating dragon kiln in Singapore. Dragon kilns are a spatial function of historic Chinese ceramics, which were endemic to Singapore until the twentieth century. The kiln functions when it is filled with small clay pieces and then fired to a high temperature. 'Firing the kiln' requires the work of many people, and requires months of preparation, as the kiln must be cleaned before and after. The dragon kiln, and the Jalan Bahar studios, were slated to be demolished and redeveloped as the nearby Nanyang Technological University expands. A coalition of artists, activists and community participants joined together in 2012–2013 to fire the kiln and to stop this development from happening through participatory art-making, thereby forcing a transformational public art encounter.

The 'Awaken the Dragon' festival was conceived by a small network of artists/curators 'over a dinner conversation' (festival curator, February 2014) held during the Chinese New Year celebrations in 2012. The idea was to connect the community to the historical arts, by filling the last operating 'dragon kiln' in Singapore with thousands of clay pieces to be fired in a celebration, and in doing so, to raise awareness of a piece of Singapore's history. As the festival's head curator explained to me in an interview,

> One of us said that we should fire the kiln. And I was the only ceramics person there, and I said, you know, yea, cool idea, but that kiln is so big – that you won't be able to fill it up … not even all the potters in Singapore! Not even if we got all the artists to come together; we would still not be able to fill up the kiln. So, I said, we would need to get thousands and thousands of people involved.
>
> (Interview, Curator, March 2013)

Thus, out of a conversation that occurred due to the alliance of a ceramics artist with artist-activists, the idea of the festival was born. What occurred next was a wide outreach to the community (Figure 3.1). Festival curators, and an army of volunteers, advertised a series of ceramic workshops which were inexpensive to attend (10 Sing. dollars) and held at various locations around Singapore, including housing estates and small cafés (such as the 'Pigeonhole', an arts café in the Chinatown/Tanjong Pagar area that closed in 2013). At these workshops, volunteer artists recounted the history of ceramics in Singapore and attendees were able to make small objects, which were then collected, numbered, photographed and brought to the dragon kiln to be fired during the festival. Eventually, all clay objects would be put on display at the National Museum of Singapore, thereby showcasing a participatory art exercise for, and by, the wider community.

I participated in one such workshop, held at the Pigeonhole café in December 2012. With whimsy and perhaps a whiff of subversion, I created a clay 'Marina Bay Sands' – Singapore's iconic (and much-maligned) waterfront hotel and casino mega structure – which I then gave to festival organisers to be fired in the kiln.

In January, on a rainy day at the tail-end of the 'Northeast Monsoon', I attended the *firing of the kiln.* A bus and taxi ride took me to Jalan Bahar, comprised of a few low-slung structures with tin roofs. In a covered courtyard was the dragon kiln, breathing fire, about 50 feet long. Several hundred people were assembled; some were helping to feed wood and to fan the flames of the kiln; others were seated at tables taking part in clay-making workshops. There were

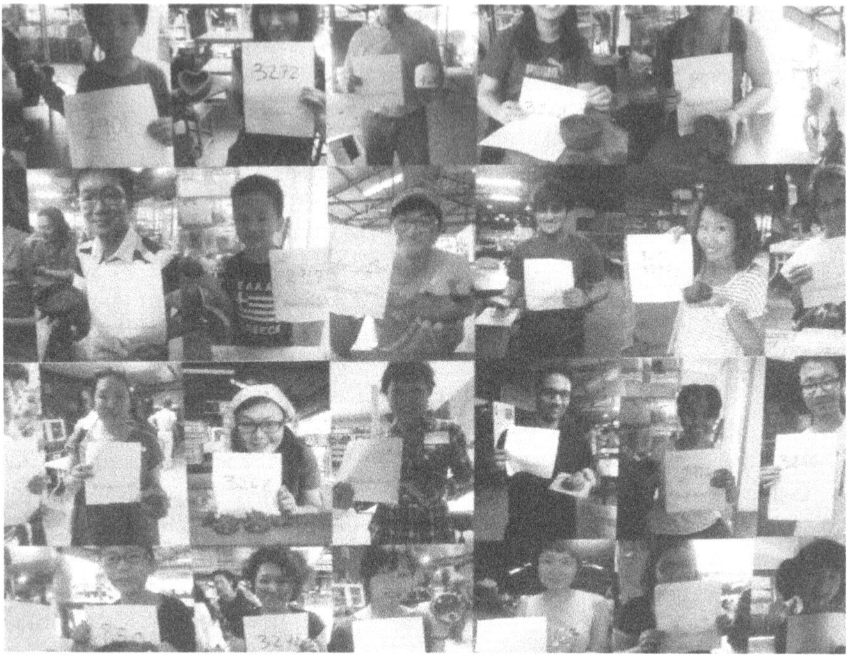

Figure 3.1 Cross-section of Singaporeans participating in the 'Awaken the Dragon' festival, 2012–2013. Credit: Jason Luger, March 2013.

food and drink stalls set up and a series of rotating presentations, where speakers (both Singaporean and foreign) described their artwork and the nature of their participation in the festival. I recognised a number of familiar faces I had encountered elsewhere in art-activist activities and events in Singapore, confirming my belief that the cultural activist network is small and tightly connected; this also confirmed my conclusion that, in this context, there is a particularly fine line between pure 'art' and 'activist-art'.

A few months later, in March 2013, I went to the festival's exhibition at the National Museum of Singapore to see my clay creation on display along with thousands of other little pieces of the city-state. By handling Singaporean clay and creating a little, imperfect piece, I had situated myself within Singapore's national story and had taken part in the *performance of nation-building*.

Feeling the land was an essential aim of the project, highly symbolic in an island-nation where land is scarce and hotly contested: Singapore's spatial restrictions have defined the nation's evolution over 50 years. As the city-state reclaims land from the sea to expand its port and financial districts, participatory art in which thousands handle Singaporean clay is a potent tactile metaphor charged with meaning. The activity allows festival participants, who span age, class, ethnic and cultural backgrounds, to handle the same Singapore clay, nation-building in a grassroots sense.

> When you are actually feeling the land – there is no – ah, there's no gray area; there is no 'but the Chinese culture in Singapore is influenced by China', or whatever, you know, or – 'Malay culture is actually part of Malaysia, and not us' – there is none of that. The land is *Singapore*. [emphasis added] It is as straightforward as it is … for us it was, we want people to hold Singapore. To make something out of the land.
>
> (Interview with artist contributing to festival, January 2013)

As a participatory grassroots activity that brought thousands of Singaporeans together, 'Awaken the Dragon' succeeded in a way that many top-down arts initiatives have failed to do (but claim to want to do). It seems to fit neatly within the stated aims of Singapore's 'Global City for the Arts' policy agenda, which calls for a 'nation of art lovers' and for 'art and culture, everywhere, everyday' (Ministry of Information, Communications and the Arts 2012).

At the same time, there is a subtle subversion in the grassroots moulding of clay objects that is not immediately evident. On the surface, this may not even be read as activism, or seem to have a cultural-activist message or theme. However, the fact that the dragon kiln was surrounded by the encroaching construction of the city-state's urban expansion, itself slated for demolition, meant that 'firing the kiln' was, in a sense, an act of rebellion against the state's development plans. A micro-resistance, perhaps, but a resistance all the same. Lefebvre (1974) wrote of 'seething forces' that lie beneath the power structure of the city: the kiln, filled with thousands of clay object, firing so bright the sky lit up – very much represents a seething (fire-breathing) force.

The result of the festival – the stoppage of the destruction of the dragon kiln and a change in the development trajectory for the site – can also be read as a *victory* of sorts; the result of art influencing (and changing) policy, and the use of art to re-think and re-conceptualise a space and place. The widespread publicity and popular success of the festival resulted not only in additional funding for the festival to become an annual, longer-term event (indicating sustainability) but also resulted in the preservation of the clay studios and the conservation of the dragon kiln (for at least a decade). The festival continues annually, at the time of writing.

The popularity of the festival gained much national attention, but just as much international attention – always important in Singapore, where state leaders are highly conscious of external reputation and often craft policy based on perceived global currents and trends. This international attention was partly due to the inclusion of several foreign artists in the festival, a result of personal friendships and working collaborations of the curatorial team.

In an interview with the festival curator, a ceramics artist who lives and works in her studio in Northeast Singapore, I prompted her to consider whether or not there was an anti-development (or any critical) message behind the festival. I asked her to consider the symbolism of the 'fire breathing' of the dragon kiln and the 'seething forces' of Lefebvre's literature. She stated this was not the case, or at least she left the possibility of a deliberate activist connection open. In her words, the festival was purely about raising awareness of an historical art form and bringing members of the community to making art that ordinarily would not:

> There was no subtle message behind it. It was just – very straightforward; we want to raise awareness, right. So how are we gonna do it. The best thing to do is to fire it [the kiln] up! Get people excited. To fire it up, we need lots of people involved. And also, we need people who know what to do. So there's the international part that came in. Of course – the urgency was driven by the fact that we don't know what is going to happen to it – it might get demolished – but that's not the – the driving force, I guess. The driving force is getting as many people as possible to know about it. The urgency though was probably driven by that. Is it? To me it's separate. Is it separate to you?
>
> (Interview, Festival curator, 30s, March 2014)

On one hand, this was a straightforward art activity with no political message. On the other hand, there was an 'urgency' to raise awareness about a piece of artistic heritage threatened by development, and by making art participatory through grassroots networks, change was induced: not only was there widespread awareness of the dragon kiln, but the development site was preserved.

Therefore, a closer reading may be necessary. The 'Awaken the Dragon' festival may not have been a deliberate act of protest, but symbolically an artistic grassroots was 'awakened'. The kiln festival could only be successful with enough participants: if the kiln did not have enough clay objects, it would not have functioned properly. This can be seen, metaphorically, to represent wider cultural activism, and indeed, civil society in Singapore. This also revisits the question posed earlier of what Singapore-style art-led activism looks like: something as subtle as community-firing of a ceramics kiln may actually be a form of context-specific activism, whether intended to be so or not. This brings to mind Chatterton and Pickerill's (2010) questions about 'activist becoming activist' and deliberate vs non-deliberate acts in the everyday. The act of showing up to a dragon kiln, far from the shopping mall or the city centre, on a rainy day, is perhaps subversive in its own way: a small act of transgression. In a city-state where, as Foucault (1976[1979]) might have observed, 'power is everywhere' – it does not take a staged protest to contest and challenge this power.

On the other hand, the festival's success can also be read as gestural, even complicit with authoritarian aims. It fits within national ideology and arts policy; there was nothing overtly political or critical about it, and in no way did it directly challenge authority. If indeed 'power is everywhere' then even firing the kiln is a reification of that power. This was, perhaps, more a sanitised representation of critical art than it was an actual critical rupture. The 'victory' of saving the dragon kiln from development may be attributed more so to the site being far enough removed from the city centre that it is not immediately required for development rather than due to the publicity generated by the arts festival; the state can afford to keep the site undeveloped for a decade (after which time it may proceed with demolition). Finally, the question remains unresolved whether the state views a critically minded, engaged artistic grassroots as a desired and necessary by-product of Singapore's re-orientation as a 'cultural city' and, therefore, whether there is actually a space for the artistic grassroots to challenge state orthodoxy and emerge as a valid critical voice. The question also remains open where exactly the Singaporean state ends and begins.

Conclusion

The 'Awaken the Dragon' festival can be read a number of ways and represents a hazy and rather ambiguous example of art-led activism: is it Zukin's (2009) conception of 'culture as control', where the performance of art becomes an emasculated, non-threatening activity, reinforcing national aims? Or does 'Awaken the Dragon' fit within the 'awakening' of Singapore's cultural grassroots – does it reflect the stickiness and cohesion of networks, coalitions and alliances that also come together for more overtly political, more radical causes?

In handling Singaporean clay in a tactile sense, festival participants simultaneously formed a common identity – building the nation – and questioned the power structure, by helping to preserve a piece of history marked for destruction in top-down planning (Figure 3.2). Using Lefebvre's (1991[1974]) theoretical framework of the spatial trilogy, the festival embodies the juxtaposition of 'represented space', 'lived space' and 'practiced space' all at once by intersecting shared artistic dreams, the performance of making art, and the authoritarian state's urban planning policy and plans. The clay is representative and represented; the imagination of the kiln's firing contrasts with the maps and images of state planning documents.

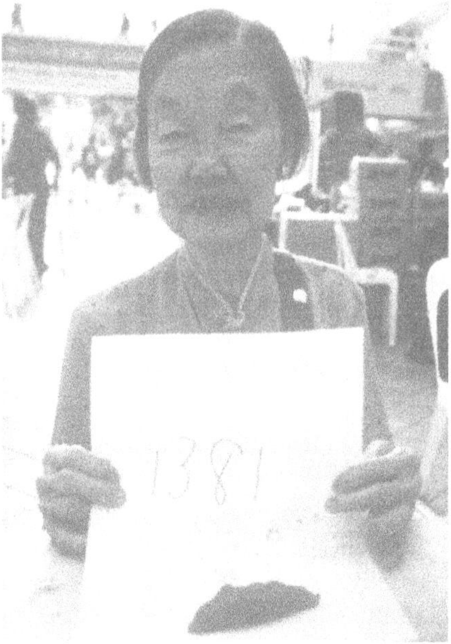

Figure 3.2 Festival Participant Number 1381, showing a personal piece of Singaporean Clay. Credit: Jason Luger, January 2013.

Foucault's notions of governmentality and the 'spaces of citizenry' (as Hartley 1992 conceptualised) are also important to consider when appraising Singapore's artist-activist encounters in the context of high-state control: 'power is everywhere' (Foucault 1976[1979]: 93). Yet, through cracks in the power structure, alternative pathways can (and are) devised – in this case, the state's recognition that urban development policy needed to change due to popular sentiment and support of preservation of the 'dragon kilns'.

One 'radical' result of the festival may be that in the future, it will be more difficult for the authoritarian state to appropriate land for development, if the 'Awaken the Dragon' model of grassroots art and place-making can be replicated. By awakening other 'dragons', more radical grassroots initiatives may come together, more quickly, to save other threatened sites and spaces that have historic or symbolic significance. As Solnit (2005) mused, resistance is often rhizomatic in nature and, like a bubbling spring, it will continue to pour forth from the ground no matter what. The important thing to take away from the festival was that a small-scale, grassroots network of artists and activists were able to come together outside and 'fire' the kiln together: this in itself can be read as an important moment in Singapore's cultural history.

The BBC noted in its report (January 2013) that 'a question still hangs over its fate: from 2015, the land it's on has been earmarked for a new eco-park. . . and unless ['Awaken the Dragon'] sparks an interest from enough Singaporeans, the spirit of the dragon . . . may be extinguished for good'. The festival's success, in the form of additional funding and the preservation of the kiln through at least 2020, indicates that when the 'dragon' of grassroots art-activism awakens, it can induce real policy shifts and alter the conversation about the production of space. However, it may be that making art can only push so far: the question remaining is whether a shared desire to shape Singaporean clay will translate into a wider grassroots desire for other, more radical outcomes.

What the Singapore case demonstrates are the creative, complicated and contradictory ways that urban space and the 'cultural commons' are used, produced and appropriated; therefore, convenient categories like 'grassroots' and 'state' are not easily demarcated or bifurcated. *Grassroots art* and *top-down art* initiatives interplay and interweave in complex ways; the public is therefore both caught in a performance which is complicit with state aims and subtly defying state aims and agendas.

If nothing else, 'Awaken the Dragon' demonstrates that the participatory *public art encounter* has the capability to generate wide publicity and motivate a broad consensus if it can connect history, space and culture in a way that is tangible, playful and fun: too often, such efforts are lost in abstraction and self-absorption and remain the domain of elites and those 'in the know'. Such a wide-reaching grassroots art effort, paradoxically, may only be possible in Singapore due to the support of the authoritarian state – and the city-state's inherent restrictions on space and artistic content. Further research is needed on the varying landscapes and diverse terrains of participatory grassroots art movements to determine which 'dragons' will awaken and which will remain sleeping.

References

Amin, A. and Thrift, N. 2013. *Arts of the Political: New Openings for the Left.* Durham, NC, and London: Duke University Press.

BBC News. 2013. *Singapore Looks to Fire up Dragon Kilns Again.* Available at: www.bbc.com/news/business-21271699 [accessed August 2015].

Buser, M., Bonura, C., Fanin, M. and Boyer, K. 2013. Cultural activism and the politics of placemaking. *City*, 17(5), 606–627.

Chang, TC. 2008. Art and soul: powerful and powerless public art in Singapore. *Environment & Planning A*, 40, 1921–1943.

Chang, TC. 2015. Housing the arts in a developmental state: Renaissance City Singapore, in T. Oakes and J. Wang (eds) *Making Cultural Cities in Asia: Mobility, Assemblage, and Aspirational Urbanism.* London: Routledge.

Chang, TC and Shirlena Huang. 2009. Geographies of everywhere and nowhere: place-(un) making in a world city. *International Development Planning Review*, 30(3), 227–247.

Chatterton, P. and Pickerill, J. 2010. Everyday activism and transitions towards post-capitalist worlds. *Transactions of the Institute of British Geographers*, NS 35, 475–490.

Chua, Beng Huat. 1997. *Political Legitimacy and Housing: Stakeholding in Singapore.* London: Routledge.

Chua, Beng Huat. 2011. Singapore as model: planning innovation, knowledge experts, in A. Ong and A. Roy (eds) *Worlding Cities: Asian Experiments and the Art of Being Global.* London: Blackwell.

Davidson, M. 2013. In the middle of a revolution... so where the hell is Stringer Bell? *City*, 17(5), 661–670.

Economist Intelligence Unit. 2015. Annual Ranking of Most Expensive Cities. Available at: www.eiu.com/home.aspx [accessed August 2015].

Foucault, M. 1979/1976. *The History of Sexuality Volume 1: An Introduction.* London: Allen.

Foucault, M. 1982. *The Subject and Power.* Berkeley, CA: Liberty Press.

Hartley, J. 1992. *The Politics of Pictures: The Creation of the Public in the Age of Popular Media.* Perth: Psychology Press.

Harvey, D. 2012. *Rebel Cities: From the Right to the City to the Urban Revolution.* London: Verso.

Katz, C. 2004. *Growing up Global: Economic Restructuring and Children's Everyday Lives.* Minneapolis, MN: University of Minnesota Press.

Kong, L. 2012. Ambitions of a global city: arts, culture and creative economy in 'post-crisis' Singapore. *International Journal of Cultural Policy*, 18(3), 279–294.

Lee, KY. 2000. *From Third World to First: The Singapore Story.* Singapore: Marshall Cavendish.

Lefebvre, H. 1974/1991. *The Production of Space.* Trans. D. Nicholson-Smith. Oxford: Basil Blackwell.

Lim, L. 2012. Constructing habitus: promoting an international arts trend at the Singapore Arts Festival. *International Journal of Cultural Policy*, 18 (3), 308–322.

Luger, J. 2015. The cultural grassroots and the authoritarian city, in Oakes and Wang (eds) *Making Cultural Cities in Asia.* London: Routledge.

Luger, J. 2016. The living and the dead in Singapore: contesting the authoritarian tourist city, in C. Colomb and J. Novy (eds) *Protest and Resistance in the Tourist City.* London: Routledge.

Mayer, M. 2009. The 'Right to the City' in the context of shifting mottos of urban social movements. *City*, 13(2), 362–374.

Ministry of Information, Communication and the Arts (MICA). 2012. *Arts and Cultural Strategic Review*, p. 26.
Ooi, Can-Seng. 2008. Reimagining Singapore as a creative nation: the politics of place branding. *Place Branding and Public Diplomacy*, 4, 287–302.
Ooi, Can-Seng. 2009. Government and creativity: arts city Singapore. *creative Industries*, 3, 44–47. [in Chinese, translated by Marina Guo, a journal of the Shanghai Academy of Social Sciences].
Ooi, Can-Seng. 2010. *Singapore's Cultural Policy: Authenticity, Regulation and Stratification*, Report on IPS Seminar, Faculty of Law, National University of Singapore, 11 August 2010.
Rae, P. 2011. Freedom of repression. *Theatre Research International*, 34(3), 310–315.
Shatkin, G. 2014. Reinterpreting the meaning of the 'Singapore Model': state capitalism and urban planning. *International Journal of Urban and Regional Research*, 38(1), 116–137.
Solnit, R. 2005. *Hope in the Dark: The Never Surrender Guide to the Changing World.* San Francisco, CA: Canongate.
Tan, K. 2008. Meritocracy and elitism in a global city: ideological shifts in Singapore. *International Political Science Review*, 29(1): 7–27.
Tan, K. 2013. Forum theater in Singapore: resistance, containment, and commodification in an advanced industrial society. *Positions*, 21(1), 189–221.
Yue, A. 2007. Hawking in the creative city: Rice Rhapsody, sexuality and the cultural politics of New Asia in Singapore. *Feminist Media Studies*, 7(4), 365–380.
Zukin, S. 2009. Destination culture: how globalization makes all cities look the same. *Center for Urban and Global Studies at Trinity College: Inaugural Working Paper Series.* 1(1).

4 The construction of post-communist ideologies and re-branding of Budapest

The case study of Statue Park Museum

Paul Clements

There have been many changes to the public sphere in Hungary since the 'silent revolution' against Soviet hegemony. One particular example refers to the reconfiguration of patrimony from the communist period which was removed from the streets of Budapest.

The government organised an architectural competition to find a suitable design for displaying this statuary using a site of donated land on the edge of the city. The winner, Ákos Eleőd, curated an exhibition entitled The Sentence on Tyranny Park (1993) which is now known as Memento or Statue Park (Pittaway 2003: 288–289). His intention was to construct a politically neutral open-air museum which functioned to reconcile the past, especially for those who were compliant in the maintenance of civil society at the time (James 1999: 304), neither deriding nor celebrating the communist era (Nash 1993).

This ingenious solution to heritage (Light 2000: 169) appears to be highly imaginative, infusing the statuary with irony as expressed by the original title, which resonated with the nature of resistance during the communist period (James 1999: 307).

Nonetheless, it is an unflattering memorial to communism and representation is ambiguous, with engagement passive or intellectualised which expresses 'cold' memory (Turai 2009). It has been criticised as a cemetery of dead statues cut adrift and positioned in a suburban wasteland (Plachy 1993), a menacing and forbidden re-inscription of meaning (Nadkarni 2003: 204).

The lack of obvious signage to guide visitors possibly exacerbates the problem of resolving a powerful history now situated out of context (James 2005: 34). So the statues invoke notions of unfinished business through improper burial procedures which devalues the former political capital of the statuary (Williams 2008: 190) and is symptomatic of Hungary's inability to clarify its messy past (Turai 2009). It offers local Hungarians and tourists a variety of interpretations.

One salient representational problem relates, albeit conversely, to the theoretical stance of Raymond Williams (1977) who revised Marxist determinism. He maintained that reductionist notions of art and culture which distil meaning to ideology are misconstrued, as they deny the primary processes of social engagement, thinking and imagination. But Hungary has been scarred by the effects of German Nazism and Russian Communism, ideological determinants that cannot be overlooked by insisting on individualised processes of audience engagement, thinking and imagination, or in this case the intended postmodern 'irony', as the statues are loaded with significance.

There are degrees to which audiences are influenced by historical and political discourses as meaning is forever changing and representation is forever vulnerable to passionate or 'hot' memory (Turai 2009) and ideological co-option. This hybrid and compromised understanding embraces artistic intention, viewer judgement, as well as wider externalities and counter-narratives, neither singular or universal in scope nor totally relative.

Diverse interpretation cannot ignore determinism and is tested by Statue Park which questions whether postmodern whimsy is able to escape history and ideology. The latter term refers to the social production of meaning, 'illusory beliefs – false ideas or their consciousness' (Williams 1977: 55), and contrasts the actuality with its representation. In Hungary today ideology is grounded in 'post-communism', referring to a range of discourses related to democracy, neoliberal capitalism and nationalism.

The trajectory of this chapter begins first with a focus on Statue Park, which comprises three related themes: situating the museum within the cultural trail of Budapest and the Hungarian story; a critical evaluation of the museum itself; and theoretical models of audience reception to help establish the complexity of meaning.

Second, the emphasis shifts to highlight notions of commodification, memory and brand in relation to Statue Park and Budapest, which comprises four related concerns: the museum signifying the consumption of communism; national governance encouraging the transformation of public space in Budapest; the convoluted reconstitution of memory; and a conceptualisation of visitors that befits the re-branding of the city.

Statue Park Museum: a partial representation of the Hungarian story

In many ways the Hungarian national story can be divided into four overlapping historical periods: pre-fascist; fascist; communist; and post-communist.

The communist era is central to Statue Park and although it is devoid of obvious signage, the *Visitor's Guide* (Réthly 2010) encourages viewers to engage with the history of the era and it offers a partial ideological and discursive framework which concurs with other sites on the 'cultural trail' in Budapest (Tourism Office of Budapest 2015). The Museum of Terror for example, situated in the former headquarters of the secret police (AVO), 'authentically' reconstructs the past by shaping the politics of fascist and communist memory. The Jewish Museum also engages directly with those darker fascist elements of the national story and the Pest Ghetto during Nazi occupation, whilst the Hammer and Sickle Tour

(Smith and Puczkó 2010) and Marxism Pub, which is adorned with symbols of communism (Turai 2009), suggests a more consumer-orientated attitude to the tourist experience. This contrasts with the pre-fascist national story which references 'safe' cultural sites, for example famous Hungarian musicians at the Museum of Musical History and visual artists at the Hungarian National Gallery, or general history, as characterised by the Budapest History Museum.

An exhibition of ambiguity

The 41 exhibits removed from the streets of Budapest in 1991 after the 'silent revolution' include 13 memorial plaques and 28 statues or busts (Light 2000: 167). The curation of the predominantly socialist realism style of statuary is polysemic with pieces juxtaposed from different chronological times (James 2005: 25), offering a range of interpretations steeped in historical, ideological, aesthetic discourses and a 'derisive nostalgia' (Williams 2008: 186). The examples are purposely chosen to reflect the ambiguity of meaning and ambivalence towards the communist period.

The entrance wall contains statues of Lenin and Marx set into a neo-classical façade which parodies socialist realism architecture (Light 2000: 168), setting the scene for the irony intended by Eleőd. The pathways, which circulate around a central figure of a five-pointed communist star of red flowers, loop back on each other forming endless figures of eight, the mathematical symbol for infinity. Another path collides with the back wall, all of which signifies the failure of communism (Figure 4.1).

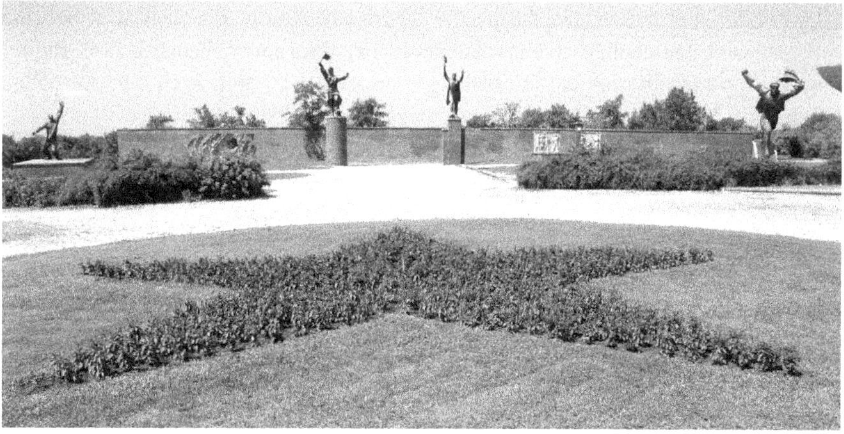

Figure 4.1 Communist Star of Red Flowers, Memento Park. Source: Paul Clements, 2010.

Outside the entrance in Witness Square (completed in 2006), there is Stalin's Boots, the reconstructed remains (the footwear) of an 8 metre bronze statue which was felled during the unsuccessful 1956 revolution (Clements 2014: 75). The original statue with accompanying grandstand was used by communist leaders to commemorate national holidays and observe military parades (Réthly 2010: 54). Stalin's Boots mocks the personality cult of Stalin and his ideology, a sentiment expressing the public mood after the 1989 'silent revolution' but hardly neutral in tone. Stalin's ghost offers much ambiguity in relation to the symbolic power of individualism and even celebrity today.

The memory of 1956 has revealed ideological divisions about what the past represents, between those Hungarians who broadly support reformist socialist aims and those conservatives and liberals who stress its anti-communist significance, as documented by the Budapest Oral History Archive (Pittaway 2003: 280). The statue for those with experience of totalitarian regimes may evoke 'hot' memory, revealing deep-seated anger and ambivalence towards the past whereas other tourist response may be colder and intellectualised, even humorous spectacle.

There are several monumental statues in Statue Park which due to their size invoke awe or a contrasting emotion of hubris. This is captured by the Monument to the Martyrs of the Counter-Revolution (Figure 4.2) cast in bronze and created in 1960 to celebrate those who were killed in defence of the Communist Party Headquarters during the 1956 uprising (Réthly 2010: 48). The notion of counter-revolution is confusing as the term has been appropriated to describe those who supported the system, not those rebels who wanted reform.

In contrast, the Béla Kun Memorial is a metal sculpture which was commissioned in 1986 by the communist regime to celebrate the 100th anniversary of the birth of Béla Kun, leader of the short-lived 1919 Hungarian Soviet Republic (Réthly 2010: 36). Kun is waving goodbye to a crowd which is a parody of those killed under the system and during the 'silent revolution' the memorial became a focal point for change (Clements 2014: 76). Moreover, because it is highly unusual for a totalitarian communist regime to satirise itself, such humour offers much ambiguity.

The range of representation and interpretations of these emblems of Hungarian communism is variable and complex. As Eleőd comments, 'a tourist for whom dictatorship is merely something they have read about, has completely different thoughts when in the park than the person with the tragic past' (cited in Réthly 2010: 6). These artefacts are serious emblems of history and ideology but they can also elicit more abstract and playful counter-narratives. This suggests the co-existence of modernist meta-narratives with more localised and mutable post-modern micro-narratives (McGuigan 2006).

The exhibition centre shows a Hungarian Secret Police (AVO) training film entitled *The Life of an Agent* (Papp 2004) which in keeping with the general tenet of the museum can be interpreted as a serious documentary about the training of secret agents in the communist era, an ironic re-interpretation of this process or as infotainment.

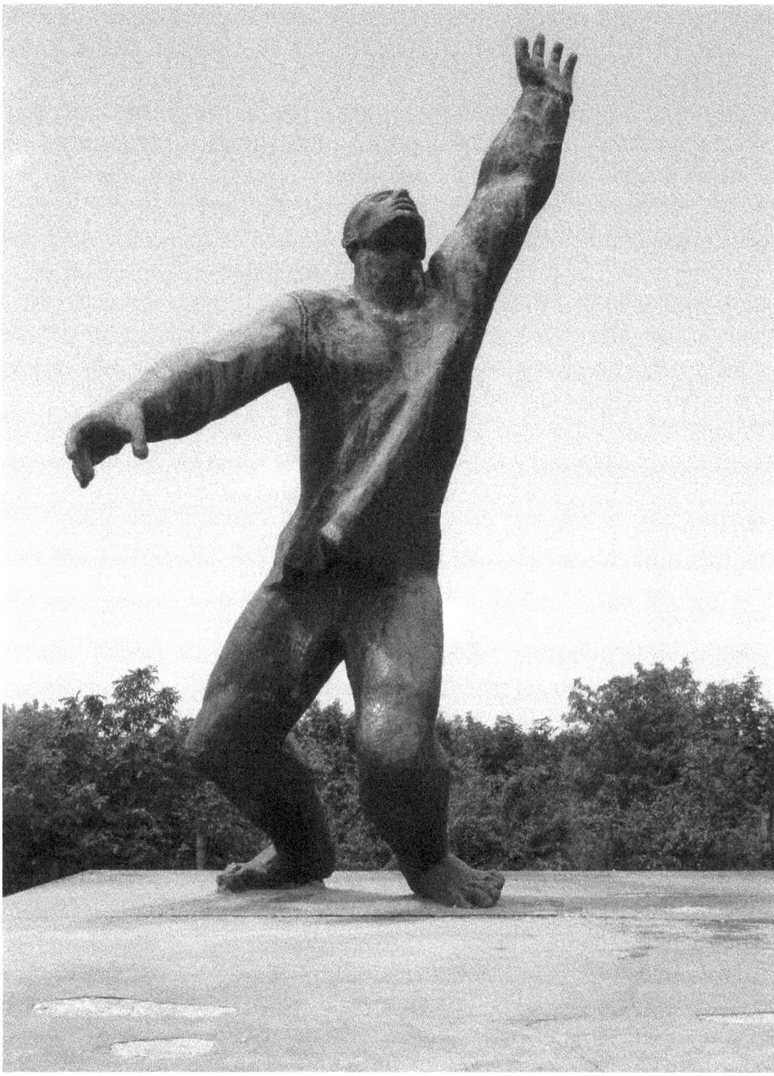

Figure 4.2 Monument to the 1956 Martyrs of the Counter-Revolution. Source: Paul Clements, 2010.

Notional audience response

Some theoretical models of audience reception may help to ground the variety of meanings offered. Klaus Jensen and Karl Rosengren (1990: 186–191) compared the five main research traditions in relation to understanding media messages which have been adapted to construct meaning positions for Statue Park and can be blended to offer a comprehensive understanding of audience response.

The Effects Tradition which is deterministic and associated with how media messages affect behaviour, acknowledges that we are influenced by the contents of culture, usually in a negative manner, which has resulted in moral panics spread by powerful gatekeepers.

It has presumed a passive human nature that is easily manipulated and very different to the Uses and Gratifications approach which, instead of taking a pejorative view of what culture may do to the audience, is concerned with the way that audiences actively use culture and find pleasure from the 'text'. Nonetheless, initially the interest in gratification concentrated on how cultural content attracted and influenced audiences allied to the effects tradition of passivity (Ruggiero 2000).

The third method relates to Literary Criticism and how the audience decodes the aesthetic or cognitive messages within specific canons of understanding. This refers to the 'art experience' and traditional 'high' culture, where the relationship between individual readers and texts is pertinent.

The Cultural Studies approach is broader in scope, structural and eclectic, as it extends an understanding of cultural representation beyond the canon, embracing broader social meanings and practices. This includes semiotic, discursive and Marxist frameworks of analysis, resistance to hegemony, as well as recognition of the co-production of meaning and the influence of class, gender and ethnicity amongst other structural variables.

Finally, Reception Analysis develops meaning to include theoretical frameworks ranging from symbolic interactionism to psychoanalysis and tries to integrate humanistic with social scientific perspectives drawing together Uses and Gratifications, Literary Criticism and Cultural Studies approaches. Its postmodern focus on the agency of individuals to determine meaning has been critical of other methods for neglecting the variety of aesthetic responses to culture corresponding to the free-play of representation. In many ways this approach struggles with the power of encoded meaning.

The Effects Tradition, Literary Criticism and Cultural Studies analyses tend to focus on the contents and actual message of the 'text' whilst the Uses and Gratifications approach makes the audience central but tends to consider micro individual aspects, with macro issues given little attention. The Reception tradition similarly treats the audience as active decoders of messages (Hall 1990), but recognises that meaning has to be interpreted in relation to the socio-cultural system and prevailing practices, thereby revealing its contextual social production. Nonetheless, there is an assumption that foregrounds audience autonomy and downplays the extent to which individual acts of interpretation are influenced by structural concerns, for example the political economy and ideology.

These approaches can be applied to Statue Park, starting with issues regarding its effect on audiences and the power of the statuary to trigger reaction. This initially refers to indoctrination and the extent to which these monuments influence citizens as they did in the Soviet period, how this has altered in relation to an empty or non-existent communist actuality, the effects of anti-communist ideology and altered reality.

With regards to the Uses and Gratifications approach, during the communist period the statuary within the city was employed by citizens as landmarks to aid urban identification and as symbols that bore political, emotional and patriotic meanings (for example supporting hegemony or gratifying people's seething discontent with Soviet-styled communism). The active individual makes choices about uses which affect interpretation and extent of engagement. Statue Park today is constructed as part of the Budapest cultural trail and the overwhelming gratification for Hungarians is one of curiosity (Réthly 2010: 63), rather than anger, pride or nostalgia.

The Literary Criticism model of analysis depends upon understanding formal interpretive codes which reference the canon and correspond to the intention of the (re)-creator, which process is complex and requires cultural education and knowledge. The construction of meaning around Cultural Studies issues and the influence of structural, ideological, historical or socio-political concerns (even issues reflecting consumer marketing and memory), will become clearer as the ideas develop in this chapter. Suffice to say meaning cannot be singular or narrow in focus. Finally, issues of context, aesthetic response and 'hot' or 'cold' memory recognise individual agency and the importance of a Reception Analysis.

Eleőd's intention was 'to summarise the individual thought-provoking elements of a historical series of paradoxes into one conceptual thought process' (cited in Réthly 2010: 7), which requires a comprehensive framework of understanding accounting for all possible representational permutations and the mutability of interpretation.

The postmodern emphasis on meaning in flux was explained theoretically by Italian philosopher Umberto Eco (1992) who referred to 'Hermetic drift', following the American semiotician Charles Peirce whose model of 'threeness' referred to unlimited semiosis (a never-ending triangulation of interpretation), whereby the meaning of a sign continually alters without any finality. Irrespective of however much an artist intends meaning, or a community of interpreters and gatekeepers reach agreement about what this is and try to finalise significance, the individual has endless agency to reconstruct significance. The artefacts in Statue Park could be an overall signifier of one meaning (anti-communism), or many meanings could be invested in each piece, all of which vie for recognition.

Eco (1992: 23) illustrated how meaning forever alters by quoting Shakespeare,

Hamlet: Do you see yonder cloud that's almost the shape of a camel?
Polonius: By the' mass, and 'tis like a camel indeed.
Hamlet: Methinks it is like a weasel.
Polonius: It is back'd like a weasel.
Hamlet: Or like a whale?
Polonius: Very like a whale.

In contrast to this endless individualised process of meaning making, the Russian theorist Valentin Vološinov (1990 [1927]) emphasised the social as the determining factor of significance. Understanding is not abstract nor continually changing outside of social context but invested within it and in turn influenced by power relations. This presents a paradigmatic understanding of fixed positions within a particular milieu which only alters in response to socio-political change. Possibly Statue Park now represents a counter-narrative of irony not in relation to the Soviet period but in relation to present social concerns over the fate of civil society and an uncertain neoliberal future (Williams 2008: 195).

This suggests a confused site of contested meaning, memory and heritage, although as Light (2000: 173) has pointed out, 'communist heritage' tourism challenges the post-communist project as it brings politics back into the equation. Also, it is a symbol of the shift to consumerism which connotes a move towards a market-determined 'freedom', symbolic of the evolution of Hungary into a capitalist economy.

Commodity, governance, memory and brand: Statue Park as commodity

The site, buildings and statuary that comprise Statue Park is a privately rented public space owned by the state and operated by a private company (Réthly 2010: 63). This mechanism of public–private partnership is akin to Privately Owned Public Spaces (POPS) which have become a feature in capitalist economies (see Clements 2014: 78–80). The ideological shift towards POPS revises and reconstructs the boundaries between public and private spheres and also the symbolic space between civil society and the state.

The statuary in Statue Park has been removed from public space notionally available to and owned by all, then reconstructed in a remote public-private space which requires payment to enter, typifying 'entrepreneurial capitalist enterprise' (Williams 2008: 189). There is communist memorabilia for sale which has been recuperated as political kitsch, objects cleansed 'of any painful or uncomfortable memories that may be invoked in their consumers' (Nadkarni 2003: 203). These retro souvenirs include: replica Red Army medals, watches, flasks, lighters, pens, postcards and posters decorated with the 'wise men' of communism; T-shirts including one entitled The Terrors with a portrait of Mao, Lenin and Stalin; model Trabant cars; a compact disc of the 'Best of Communism' music and 'Party House Mix' of communist-era recordings; and the DVD, *The Life of an Agent*. Finally, the visitor can purchase 'the last breath of communism, sealed in a tin'. The sale of communist memorabilia is dripping with kitsch, gimmickry and ironic humour which is intended to titillate consumers.

The *Visitor's Guide* (Réthly 2010) has four pages promoting sales from the Red Star Store just in case the visitor fails to see the merchandise for sale at the entrance/exit. The official website (www.mementopark.hu/) offers other merchandising including an officially stamped Soviet Passport (written in either Russian or English). It also details synergistic ties with 'Partners in Tourism' and 'Service Partners' including a website of hotels and hostels, a travel agency, a Budapest tourist information site and a digital technology company, amongst other businesses. Statue Park is franchised on a self-sustaining basis, a business which relies on entry fees and souvenir sales.

This consumption of communism is hegemonic and invokes Jim McGuigan's (2011: 11) notion of 'cool capitalism' and '*the incorporation of disaffection into capitalism itself*', which has added symbolic power within this specific context as it has consumed its old enemy. Chris Rojek (2003: 114–115) summarised the process of engineering consent as operating on a number of levels which are purposely obscured. Hegemony frames and positions what difference and resistance are, but it is erroneous to perceive the governance of normalisation as an intentional expunging of different ideologies as this fails to comprehend how, in this case, the process manipulates and co-opts the signs of communism in order to recuperate them. These are now being sold off as kitsch in Statue Park, driven by a market philosophy which has altered the meaning of the artefacts. There appears to be little scrutiny or debate about the past beyond the information presented in the *Visitors Guide*, which is critical of the era employing an ironic tone in parts.

Possibly it is just a matter of time before this museum and similar systems of renting public space are sold off. This is reminiscent of the privatisation of services and state assets in the UK which was instigated by the right-wing neoliberal government of Margaret Thatcher in the late 1970s. Richard Sennett (1986 [1974]) critiqued the privatisation of the public sphere for constructing an imbalance of human experience which has downgraded the public realm of political life in favour of private psychological self-examination. In effect, we have become so self-absorbed and self-concerned that we take little interest in the world and public life outside our own life, agendas and social networks, a process exacerbated by the fragmentary effects of capitalism and technology. We project our psychological frameworks onto the world rather than interpreting it outside of them, and look inwards rather than outwards, which corrupts the ideal of citizenship and a public realm where people are able to voice their socio-economic and political needs, differences and concerns.

Viktor Orbán's governance and the commodification of Budapest

Within capitalist economies there is a tension between the market (economic liberalism) and democracy, free speech and human rights (socio-cultural liberalism), whereas in Hungary during the communist era the struggle for socio-cultural liberalism was at odds with an oppressive state. Notwithstanding this, the post-communist right-wing Fidesz government of Viktor Orbán first elected in 1990 appears to be operating by employing both the power of global capitalism and the state to deny citizen rights, a system which combines neoliberal ideology with centralised control and possibly better configures the realpolitik of neoliberal practice.

The process of recuperating the communist past through 'cool' capitalism has not been aided by the faltering state of the political economy. During this period there has been greater income inequality, declining living standards and increased unemployment (Onaran 2009), an accumulation of national debt and ongoing repayments of large International Monetary Fund loans. The consequence has been that many state subsidies have been removed, creating much poverty which is a new feature of Hungarian society (Fabry 2009). Although the Hungarian Socialist Party has gained power several times, it has pursued a similar political line just as other socialist parties in Western Europe, all of whom have promoted degrees of neoliberalism when elected.

The ideology of neoliberalism is a political and economic configuration which promotes the hypothetical notion that the optimum advancement of human beings within society requires individual entrepreneurial freedoms set within an institutional framework, the promotion of private property rights, 'free' markets and 'free' trade (Harvey 2005: 2). This global 'utopia of the pure market' (Bourdieu 1998: 96) requires financial deregulation and the deconstruction of collective structures where the state is reduced to the role of ensuring the best conditions for markets to operate in. Through POPS it reconfigures the partnership between private and public domains, fashioning a consumerist model of civil society. This privileges commercial interests and is far from ideal as it is immersed in an ideology which is elitist, obfuscates neutrality and objective 'public' representation crucial for a democratic public sphere.

The transformation of physical space in Budapest highlights in a practical manner its infiltration by market capitalism with a boom in shopping malls and business parks, and only fragmented opposition offered by Trade Unionists and environmentalists (Szalai 1998 cited in Bodnár 2001). There has been much structural change since the 'silent revolution' and the liberalisation of the once state-owned real estate market which privatisation of housing has proceeded faster in Budapest than other ex-communist bloc countries (Grime 1999). This is not totally surprising as Hungary in the past was recognised as one of the most advanced consumer societies within the Communist Bloc when there were debates concerning false needs undergirding commoditisation and technical fetishism (Crowley 2003: 279).

Today in the city centre the more affluent consumers and tourists have displaced local shoppers at retail centres, which have been constructed and purchased by multinational companies (Timár 2001), and there has been a 15–20 per cent increase of floor space per year during the 1990s (Nagy 2001: 342). Moreover, the rolling back of state responsibilities has adversely affected essential welfare services, health and education. Keith Grime (1999: 40) has maintained that the populace of Budapest has had reservations, especially about the large-scale privatisations of housing and buildings more generally, with 65 per cent surveyed dissatisfied with the inequalities of distribution. Subsequently, the Orbán/Fidesz government has sold off prime agricultural land leased from the state including organic farms in Kishantos which were once thriving producers of food, to 'oligarchs and businessmen' (Traynor 2014), creating a crony state.

The personality cult associated with Orbán is reminiscent of Stalinism (or Putinism), and there are questions regarding the extent to which the democratic gains made since the 'silent revolution' have been eroded. The highly centralised Fidesz government has removed checks and balances to its own power with the most vulnerable social groups becoming peripheral. There has been a marginalisation of democracy and much disillusionment amongst the Hungarian population, with a mere 60.2 per cent turnout in the 2014 election in which the greatest increase of the vote was for the far-right Jobbik party (Mudde 2014). The government is unpopular with allegations of corruption, street protests against new taxes and criticism about cosying up to Russian President Vladimir Putin over energy deals.

Post-communist cities have undergone transformation as they have adapted to changing political and socio-economic realities. The state and local government, starved of resources, have had to engage in partnerships with corporate wealth reminiscent of Anglo-American development (Sailer-Fliege 1999). But the scenario is patchy as corporate capitalism has been attracted to those areas in cities which either best promote their organisations or are highly visible and desirable, therefore helping to market their products and ideology. Hence it is unsurprising that there are 'attractive' areas of Budapest which have been transformed into capitalist hubs whilst other 'uglier' sectors of the urban landscape resemble 'frozen mirrors of socialism' (Sýkora and Bouzarovski 2012: 44), creating an uneven city.

Public space and the POPS containing these capitalist hubs, present opportunities for branded heritage sites and the display of corporate brands for commercial interests, which visually pollutes and dumbs down the urban aesthetic. Naomi Klein (1999) created the battle cry of 'No Logo' as a riposte to the branding of the world which has reified space and language underpinned by marketing, business and management jargon. It has falsified knowledge by employing, 'imagery to equate products with positive cultural or social experience . . . [which] seeks to take these associations out of the representational realm and make them a lived reality' (1999: 29). Correspondingly, privately rented public spaces are not disinterested arenas for debate, and Statue Park encourages the consumption of communism by employing 'cool' ironic signification to re-inscribe the artefacts with superficial postmodern meanings that focus on the present, which obfuscates history.

The 'ideal' public sphere promotes citizenship and democracy which gives voice to all rather than to privilege 'brand bullies' who dominate the visual environment and media in order to develop people as consumers. Habermas (1989 [1962]) referred to this type of realm as a pseudo public sphere of superficial debate which denies rational dialogue, thereby downgrading the importance of democratic rights. Neoliberal discourse dismantles and co-opts the concept of 'public good' into a material concern for open markets and competition (Murdock 2004), rather than any interest in developing a fully rounded, diverse and educated citizenship.

Cultural tourism aids this process as it attempts to construct individual and group identities through particular lifestyles and consumption practices (Britton 1991), a process that has gathered speed in Hungary.

The reconstitution of memory

Memory can be configured to re-frame the present (Hewison 1987), where heritage is recuperated nostalgically as a homogenous imaginary to avoid diverse, critical, complex and negative viewpoints. For Statue Park the reconstitution of memory is even more convoluted because the statues have been removed from their intended sites, with meaning de-contextualised then re-contextualised.

Postmodernism suggests a different representation of reality than modernist discourse. For example, one of 'irony' which presents superficial floating signification, rather than 'shame' which offers depth and an ethical debate in order to engage with history (Clements 2014: 75). Nonetheless, this 'irony' may engage with criticality and ambiguity which, together with 'Hermetic drift', can challenge rigid structures and therefore strengthen democratic debate (Willet 2008).

Reconfigured heritage which encourages counter-hegemonic memory (memory-as-resistance) may prevent an evaluation of communist ideology, as it readily creates amnesia regarding the past and overlooks the existence of alternative positions (Esbenshade 1995: 75). Narratives of the past operate on many levels, including top-down official versions and bottom-up local understandings. Moreover, there is ambiguity regarding memory and heritage in Eastern and Central Europe as individual responses to the falsified history and collective memory, 'is alternatively unreliable and all too reliable in dredging up a highly compromised past' (1995: 76–77).

Charles Maier (cited in Turai 2009) has critiqued the association of 'hot' memory stirred up by Nazi atrocities with 'cold' memory allied to communist crimes, as though fascism is the catalyst for active memory. This is problematic for Hungarians as there were a range of victims from the communist period; those killed, imprisoned and unfairly discriminated against, which possibly includes all those who were involved in the experience (Toth 2011: 54). A 'cold' or passive relation to memory is closed off, 'not kept open, not worked through' (Turai 2009: 99), which contrasts with active memory surrounding the Holocaust and other atrocities which elicit moral outrage against humanity and demand closure. Victims of the regime may well have passionate 'hot' issues to contend with which refer to personal injury and anger against gross injustice.

For Statue Park the 'cold' ironic purpose intended by its creator does not match up to the need to engage with the past and ensure that such a Stalinist style of occupation does not reoccur.

In Hungary, there has been a struggle between nostalgia for the communist system and an anti- or post-communist consensus often driven by renewed nationalism, democratic and capitalist ideologies. The communist era was statist and relatively homogeneous in contrast with post-communism which although centralised appears more heterogeneous, leaving citizens ambivalent towards the notion of a state (James 1999: 301). Heritage is under pressure from commerce, adding to the confusion between the public good and the safeguarding of history which has been sterilised and reconfigured, affecting those seeking authentic experience (Dallen and Boyd 2003: 244–253).

There has been much shift-shaping of material manifestations of the communist regime in Budapest, with street names for example reverting to pre-communist terminology (see Clements 2014: 81). This practice may not resolve the ambivalence Hungarians feel towards this era, including both anger and nostalgia, and it contrasts with other ex-Iron Curtain countries. In East Berlin, for example, street names of German communist revolutionaries have been retained (including Karl-Marx-Allee, Karl-Leibknecht-Strasse and Rosa-Luxemburg-Platz).

Personal experience is not the same as official memory and Piotr Piotrowski (2012) has revisited the ambivalence of Hungarians towards the statuary from the communist era in order to unravel the process of forgetting. Unlike the removal of monuments in Budapest for Statue Park, which received little opposition from the cultural establishment in Hungary, the artists Bálint Havis and András Gálik (better known as Little Warsaw) created a storm. With permission from the local government of Hódmezővósárhely, they removed the statue of János Szántó Kovacs (the leader of the Hungarian Peasant Rebellion) which was created by Józsof Somagyi in 1965. This was displayed in an exhibition at the Stedelijk Museum (2004), Amsterdam, entitled *Time and Again.*

In response the Academy of Art in Hungary condemned this in a signed petition as a denial of the authorial rights of Somagyi's daughter (an art critic). The citizens of the town who previously seemed disinterested in its temporary removal now recognised its absence as an attack on the town's identity. Moreover, there is a political point being made about a post-communist Hungary, and the need to champion the rights of individuals over the collective (influential daughter above town council). Piotrowski (2012: 172) maintained that

> no one protests when 'openly' communist monuments are removed and placed in 'monument parks' or simply destroyed, when it comes to public sculptures with less clearly indicated communist status (depicting other leaders, using other styles) the issue of identity of a place created within structures of memory appears.

But although he argued that the statuary from the communist period serves the function of forgetting rather than remembering (an amnesia only disturbed when people notice absence), there are more obvious ideological issues concerning history and the promotion of a post-communist discourse. Interestingly the Kovacs statue is of a Hungarian hero, not a communist leader, and the style is neither socialist realism nor heroic. Possibly there is greater association of the monumental realist style with the communist past as represented by the *Monument to the 1956 Martyrs of the Counter-Revolution*, which befits hubris.

A re-branded city

There have been roughly 60,000–70,000 visitors to Statue Park per year of which roughly half are Hungarian whilst others tend to be tourists from outside Eastern and Central Europe (Réthly 2010: 63). It is marketed to tourists (and locals) through the government-sponsored Hungarian National Tourist Office and Tourist Office of Budapest (TOB) which is funded locally (Smith and Puczkó 2010: 292), and both have been key to constructing the 'cultural trail' for a re-branded city.

It is important to iterate the problems with the etymology of the term 'brand' as it is derived from the marking of people rather than marketing of ideas. For example, in the UK during Tudor times (sixteenth century) the deserving poor wore badges which were associated with shame (Hindle 2004), and slaves throughout time have been branded accordingly, defining their commodification as chattels. The negative manifestation of the term is further highlighted in this context as the Nazis tattooed Hungarian Jews in concentration camps with numbers on their forearms and prisoners were branded in Soviet gulags with hot irons which equates to branding livestock as a sign of ownership. Also during the communist purges, thousands of Hungarians were 'branded' as spies and saboteurs.

This negativity reappears when applied to the commodification of culture, a practice which for the tourist industry embraces place-branding and the application of commercial marketing processes to geographic locations. This collection of identities signifying place may not complement each other and these may be manipulated in order to shift perceptions (Anholt 2002: 232). The branding of a post-communist Budapest becomes a political game of appropriating and reconfiguring signifiers rather than complete reinvention, in order to manage perception and refresh the narrative (Julier 2005: 885, 872). It concurs with the reinvention of national identities through the reconstruction of urbanity and cultural traditions (Hobsbawm and Ranger 1983). Marketing strategies attempt to garner the emotional effect of heritage and aesthetic markers (often buildings, statuary or other visual signifiers). For post-communist Budapest this references 'high' cultural capital offered by the 'cultural trail' including the cool irony of Statue Park, which helps 'brand' it as the 'Paris of the East'. Not unsurprisingly this includes a concentration on 'safe' pre-fascist built heritage attractions which reconstruct the 'Golden Age' of the 1890s that emphasise the Danube River (Smith and Puczkó 2010: 292–293).

Perception studies of budget airline travellers undertaken in 2005 suggest a muddy picture regarding the image of Budapest as a cultural centre with marked national differences regarding the gratification for visiting by different nationality groups:

> visitors from Northern Europe (Denmark, Norway, Sweden) were motivated mainly by the lower . . . prices. The French and Swiss came to visit cultural sites and events. The British were driven by a curiosity factor. Many Germans (especially East Germans) were keen to return to a place they had visited in the socialist era and for which they were nostalgic . . . cheap airline tickets, however were unsurprisingly the main reason for budget airline passengers to come to Budapest.
>
> (Hungarian National Tourist Office cited in Smith and Puczkó 2010: 292)

Notwithstanding the specifics of the 'cultural trail', Budapest can be compared with other European destinations (notably Prague). The rationale of budget travel homogenises and commodifies urban experience as often the tourists are less interested in the unique selling points of the city than the cost of the holiday, entertainment and commodities.

Consequently, the TOB (created in 1996), has promoted a vibrant and dynamic young image of a unique spa and gastronomic centre where the cultural tourist is constructed to spend money. For example, the Budapest Card (Tourism Office of Budapest 2015) offers a cheap package for tourists to gain entry to a range of cultural institutions, guided tours, baths, transport, restaurants and clubs, which commoditisation of experience is the rubric of capitalism.

In 2004 the Tourist Development Strategy emphasised five values to promote a vibrant city that encapsulate the Budapest experience: an atmospheric, entertaining, pampering, spectacular and dynamic city. Whilst the first four concern heritage, culture and the Danube River, the latter category recognises the development of the city as a business centre (cited in Smith and Puczkó 2010: 293). This conflates and imbues the notion of 'dynamism' with entrepreneurial practices which suggests a 'natural' relationship and synergy with the other values, a good example of ideological encryption.

Nonetheless, cultural tourists traditionally have wanted emphatic experiences and to immerse themselves in cultural knowledge (Huyssen 1995), heterogeneous experience rather than 'packaged' homogeneity reminiscent of mass tourism. Here engagement with the 'cultural trail' corresponds to the Uses and Gratifications, Literary Criticism and Reception Studies models of what the audience does with culture, the decoding of aesthetic signs and individual perception, which is varied.

This may include a search for dramatic cultural encounters or *in extremis* 'trauma tourism' which encrypts 'hot' memories of the past. For example, The House of Terror in Budapest; the preserved Auschwitz concentration camp in Poland; the Museum of Terror which lies on the site of the old Gestapo headquarters in Berlin; and Grutas Park (or Stalin's World) in Lithuania which contains 86 Soviet-era statues within a reconstructed labour camp with barbed wire and military music from the era (Williams 2008: 187–188). All of these cultural spaces demand 'hot' emotional responses which contrasts with the 'cool' irony of Statue Park.

Smith and Puczkó (2010) suggest that there have been too many attempts at branding Budapest, creating confusion, partly due to its tricky past and memories associated which disrupts the (re)construction of a post-communist consensus. There has been a greater focus on the commodification of experience that a cultural studies, political economy and even effects approach helps to unravel, which enhances an understanding of audience reception. It offers a broader and deeper configuration of meaning, reflecting on what culture does to the audience that accounts for the negative effects of memory, the branding and marketing processes and other responses to socio-political change.

Statue Park personifies ambivalence and is a touchstone of the changing times, emplacing the difficulties of relating to a troubled and messy past. Interestingly, this confusion was symbolised by Eleőd who justified his architectural design by suggesting that he did not want the artefacts to be anti-propaganda, which is 'a continuation of dictatorship mentality' (cited in Réthly 2010: 6), as though they can be non-ideological, which within this context is baffling.

The 'Budapest experience' resonates globally as there are issues regarding the extent to which cultural tourism is configured to reflect capitalist ideologies, a functionalism that manipulates and obfuscates engagement with national or local history and diverse identity, creating readily identifiable brands which operate to deliver consumerism. Nowhere is this homogeneity better symbolised than by the synergistic TOB promotion of an outlet of McDonalds in Budapest, situated in a building designed by Alexandre Eiffel (whose company constructed the railway terminal at Budapest-Nyugati Pályaudvar), and is best known for the iconic Eiffel tower in Paris (Smith and Puczkó 2010: 193). In May 2014 the 100th McDonalds restaurant opened in Hungary (see www.kisalfold.hu/gyori_ hirek/gyorben_ nyilik_a_mcdonald8217s_100_magyar_etterme/2112154/) to consolidate the commoditisation of the public sphere and the hegemony of neoliberal capitalism in post-communist Budapest.

Conclusion

Statue Park offers complex representations of the communist past. It is over 20 years old and the irony originally intended is open for re-interpretation today in light of post-communist ideologies. It highlights the problematic relationship between postmodern micro-narratives that focus on individual interpretation, and social determinants of history and ideology related to the political economy with associated macro-narratives which have 'effects'. Postmodernism represented by cool cosmopolitan 'irony' may suggest a socially liberal 'openness' of interpretation, but it overlooks the ideological encryption of these values within a 'free' market philosophy of economic liberalism which corresponds to the commodification of Budapest.

In terms of interpretation, the article situates two contrasting theoretical problems associated with meaning making. This concerns the degree to which audiences have agency and playfully create sense from cultural signifiers through a process of 'Hermetic drift', or the extent to which fixed meanings are constructed within specific social contexts encouraged by ideological signifiers. The 'free' market promotes an ideology of agency and difference, an individualism and focus on consumerism which obfuscates the central role of history. Furthermore, the artefacts in Statue Park are in limbo and without resolution which postmodern meaning denies any clear position, hence the ambivalence of response by visitors. Reaction to the Soviet era in Hungary may concern a range of emotions which may be latent and only recognised when the landscape is altered, as highlighted by the 'Little Warsaw' appropriation of statuary.

Statue Park signifies how Budapest has been re-branded as a tourist hub similarly underpinned by post-communist ideologies, a recuperation of the past which offers opportunities for entrepreneurial enterprise. There still remains an all-powerful Hungarian government with a prime minister who sells off state assets to his cronies which reveals the political reality of neo-liberalism in contrast to its representation as an individualised ideology steeped in democracy, 'freedom' and notions of meritocracy. This is in line with the commoditisation and revision of the public sphere, but is especially significant in Hungary where a new configuration of politicians, oligarchs and global corporations have replaced both fascist and communist incarnations of a corrupt state. Irony appears to be a very sustainable discourse as historically none of these systems appears to have delivered for Hungarians.

Possibly the most ironic statement was made by the director of Statue Park who suggested that Marx and Engels' infamous *Communist Manifesto* was a 'valid diagnostic analysis of capitalism' and an 'incredibly accurate insight into the future'. Then he cited from it, suggesting that '[t]he bourgeoisie has, through its exploitation of the world market, given a cosmopolitan character to production and consumption in every country' (Réthly 2010: 17), although he failed to mention the complicity of Statue Park in this process.

References

Anholt, S. 2002. Foreword. *Journal of Brand Management*, 9(4/5), 228–240.

Bodnár, J. 2001. *Fin de Millénaire Budapest: Metamorphoses of Urban Life*. Minneapolis, MN: University of Minnesota Press.

Bourdieu, P. 1998. *Acts of Resistance against the New Myths of Our Time*. Cambridge: Polity.

Britton, S. 1991. Tourism, capital, and place: towards a critical geography of tourism. *Environment and Planning D: Society and Space*, 9(4), 451–478

Clements, P. 2014. The consumption of communism: changing representations of Statue Park Museum in Budapest. *Art & the Public Sphere*, 2(1–3), 73–86.

Crowley, D. 2003. Making the socialist home in postwar Eastern Europe, in *Globalization and Europe*, edited by M. Pittaway. Milton Keynes: Open University, 249–293.

Dallen, T. and Boyd, S. 2003. *Heritage Tourism*. Harlow: Prentice Hall.

Eco, U. 1992. Unlimited semiosis and drift, in *The Limits of Interpretation*. Bloomington, IN: Indiana University Press, 23–43.

Esbenshade, R. 1995. Remembering to forget: memory, history, national identity in East-Central Europe. *Representations*, 49, 72–96.

Fabry, A. 2009. End of the liberal dream: Hungary since 1989. [Online: *International Socialism*, 124]. Available at: www.isj.org.uk [accessed: 24 September 2011].

Grime, K. 1999. The role of privatisation in post socialist urban transition: Budapest, Krakow, Prague and Warsaw. *GeoJournal*, 49, 35–42.

Habermas, J. 1989 [1962]. *The Structural Transformation of the Public Sphere*. Cambridge: Polity.

Hall, S. 1990 [1980]. Encoding/decoding, in *Culture, Media, Language; Working Papers in Cultural Studies 1972–9*, edited by S. Hall, D. Hobson, A. Lowe and P. Willis. London: Routledge, 128–38.

Harvey, D. 2005. *A Brief History of Neoliberalism*. Oxford: Oxford University Press.

Hewison, R. 1987. *The Heritage Industry. Britain in a Climate of Decline*. London: Methuen.

Hindle, S. 2004. Dependency, shame and belonging. Badging the deserving poor, c. 1550–1750. *Cultural and Social History*, 1(1), 6–35.

Hobsbawm, E. and Ranger, T. 1983. *The Invention of Tradition*. Cambridge: Cambridge University Press.

Huyssen, A. 1995. *Twilight Memories*. London: Routledge.
James, B. 1999. Fencing in the past: Budapest's Statue Park Museum. *Media, Culture and Society*, 21(3), 291–311.
James, B. 2005. *Imagining Postcommunism*. College Station, TX: Texas A & M University Press.
Jensen, K. and Rosengren, K. 1990. Five traditions in search of the audience. *European Journal of Communication*, 5(2/3), 207–238.
Julier, G. 2005. Urban designscapes and the production of aesthetic consent. *Urban Studies*, 42(5/6), 869–887.
Klein, N. 1999. *No Logo*. London: HarperCollins.
Light, D. 2000. Gazing on communism: heritage tourism and post-communist identities in Germany, Hungary and Romania. *Tourism Geographies*, 2(2), 157–176.
McGuigan, J. 2006. *Modernity and Postmodern Culture*. Maidenhead: Open University Press.
McGuigan, J. 2011. From cultural populism to cool capitalism. *Art & the Public Sphere*, 1(1), 7–18.
Mudde, C. 2014. The 2014 Hungarian parliamentary elections, or how to craft a constitutional majority [Online: *The Washington Post*, 14 April]. Available at: www.washingtonpost.com/blogs/monkey-cage/wp/2014/04/14/the-2014-hungarian-parliamentary-elections-or-how-to-craft-a-constitutional-majority/ [accessed: 18 June 2015].
Murdock, G. 2004. Past the posts re-thinking change, retrieving critique. *European Journal of Communication*, 19(1), 19–38.
Nadkarni, M. 2003. The death of socialism and the afterlife of its monuments. Making and marketing the past in Budapest's Statue Park Museum, in *Contested Pasts: The Politics of Memory*, edited by K. Hodgson and S. Radstone. London: Routledge, 193–207.
Nagy, E. 2001. Winners and losers in the transformation of city centre retailing in East Central Europe. *European Urban and Regional Studies*, 8(4), 340–348.
Nash, E. 1993. Communism: the theme park? *The Independent on Sunday*, 25 September, 40–41.
Onaran, Ö. 2009. Global crisis and the policy reaction in Western and Eastern European Union. Can policy save capitalism from itself? [Online: The Berlin Institute for International Political Economy]. Available at: www.ipe-berlin.org/uploads/media/Ozlem_Crisis_and_policy_in_EU.pdf [accessed: 24 February 2015].
Papp, G. 2004. *The Life of an Agent* [Edited highlights of Hungarian government training film]. Budapest, Hungary: Bologna Film.
Piotrowski, P. 2012. *Art and Democracy in Post-Communist Europe*. London: Reaktion Books.
Pittaway, M. 2003. Dealing with dictatorship: socialism and the sites of memory in contemporary Hungary, in *War, Culture and Memory*, edited by C. Emsley. Milton Keynes: The Open University, 267–308.
Plachy, S. 1993. Graveyard of the statues: communist heroes in perspective. *The New York Times Magazine*, 2 May, 46.
Réthly, Á. 2010. *In the Shadow of Stalin's Boots, Visitors Guide to Memento Park*. Budapest, Hungary: Private Planet.
Rojek, C. 2003. *Stuart Hall*. Cambridge: Polity.
Ruggiero, T. 2000. Uses and Gratifications theory in the 21st Century. *Mass Communication and Society*, 3(1), 3–37.
Sailer-Fliege, U. 1999. Characteristics of post-socialist urban transformation in East Central Europe. *GeoJournal*, 49(1), 7–16.

Sennett, R. 1986 [1974]. *The Fall of Public Man*. London: Faber and Faber.
Smith, M. and Puczkó, L. 2010. Out with the old, in with the new? Twenty years of post-socialist marketing in Budapest. *Journal of Town and City Management*, 1(3), 288–299.
Sýkora, L. and Bouzarovski, S. 2012. Multiple transformations: conceptualising the post-communist urban transition. *Urban Studies*, 49(1), 43–60.
Timár, J. 2001. Introduction: uneven development in Hungary. *European Urban and Regional Studies*, 8(4), 319–320.
Toth, M. 2011. *Symbols in Transition: Public Places, Reparatory Justice, and the* Statue Park [Unpublished thesis]. Central European University, Budapest.
Tourism Office of Budapest 2015. [Online: Budapest Card Booklet]. Available at: http://budapest-card.com/en/ [accessed: 22 March 2015].
Traynor, I. 2014. Budapest autumn? Hollowing out democracy on the edge of Europe. *The Guardian*, 30 October, 17.
Turai, H. 2009. Past unmastered hot and cold memory in Hungary. *Third Text*, 23(1), 97–106.
Vološinov, V. 1990 [1927]. Discourse in life and discourse in art, in *Freudianism, A Marxist Critique.* Translated from Russian by I. Titunik and edited in collaboration by N. Buss, Bloomington and Indianapolis: Indiana University Press, 93–116.
Willet, C. 2008. *Irony in the Age of Empire: Comic Perspectives in Democracy and Freedom*. Bloomington, IN: Indiana University Press.
Williams, P. 2008. The afterlife of communist statuary: Hungary's Szoborpark and Lithuania's Grutas Park. *Forum for Modern Languages*, 44(2), 185–198.
Williams, R. 1977. *Marxism and Literature*. Oxford: Oxford University Press.

5 Sustainable influences of public art

A view on cultural capital and environmental impact

Cameron Cartiere and Ashley Guindon

Introduction

It is Mother's Day, May 2014, and a group of artists are creating a mural on the street. Some are students, others are established practitioners, and this is day two of their volunteer efforts. The first morning was damp and chilly with periodic bouts of mist that forced the team to dash into various cars to continue working as they cut out templates to lay down the initial mural design. Day two brought clear skies, but also the heat as the sun quickly heated up the macadam. As the mural began to take shape, a swirling array of salmon emerged from the street and the visual transformation garnered countless 'thumbs up' from families walking into the nearby restaurant for dim sum to celebrate their respective matriarchs. Some would stop to ask what was happening. Others would simply watch for a moment or two, smile and then move on. The occasional question would come from spectators wondering why these young people were not with their own mothers on this special day. We would explain that these students and artists lived away from their families and that they purposely chose to work on Sunday, so that other members of the team, who worked the day before, could in fact be with their mothers. This would usually generate two thumbs up and a comment along the lines of how proud their mothers must be.

The artists have been working for hours on their hands and knees when a passerby shouts out, 'That better not be my tax dollars at work.' Such discouraging comments are, unfortunately, not uncommon, and it has been our experience over years of working in the public realm that while one may receive 50 positive comments from the public, it is that one negative comment that can devastate the morale of a group of young people working under the hot noon sun.

However, in this case, our unsuspecting passerby received a shock when two students leapt up and said 'Well, actually sir, for every dollar invested in a project like this, the community gets huge rewards back, including cost-saving through graffiti reduction, increase in property values, and enhanced community spirit!' To which our critic responded by hurrying down the street shaking his head. We have trained our research team and students to respond to this type of misconception in a positive way. Ironically, the particular street mural described above was entirely supported by grants and in kind donations, not municipal taxes.

This interaction begs the question: how do you measure and prove the impact of public art on a community? There are so many factors at play in the social evolution and transformation of the urban fabric. Each community is different and environmental challenges, financial concerns and underlying civic issues can all present variable impacts.

> [T]he value of public art cannot be derived solely from an assessment of completed works, but must include its impact as a significant economic driver, its collaborative and complementary value as a component of public spaces, and its social value as a means through which to express, reflect and enjoy our city.
>
> (City of Calgary 2004: 1)

Public Art is continually promoted as a means of improving community, engaging industries and contributing to neighbourhood beautification (particularly within an urban context). As a result, public agencies are under continual pressure to define just how such improvements can be substantiated. However, there is limited qualitative data on the impact of public art (Selwood 1995, Senie 2002, Zuidervaart 2011) and there is a considerable lack of quantitative data. Communities wishing to increase their use of public art will often find it difficult to access a significant range of tangible facts about what public art can do for them in terms of developing industry partnerships, increasing community engagement, improving the urban landscape, addressing environmental concerns and/or fostering social and cultural understanding.

This chapter explores some of the assumptions regarding the long-term effects of public art in the urban environment and examines the development of a multidisciplinary public art research project in Vancouver, British Columbia that was specifically designed to examine the question of how to gauge the social impact of art on community. Responding to such a complex challenge required a coordinated, comparative strategy that evaluated different models of development for artworks and assessed the range of responses to artworks once they are in place. The Vancouver study attempted to deploy a combination of research methods to gain a multi-dimensional assessment of how specific public art projects within the community related to a wider range of effects.

Locating the public in public art

As a practice, public art is arguably one of the most contentious disciplines an artist can undertake. Historically, public art was primarily associated with public sculpture, so to embrace the practice one was generally considered a sculptor. But now that the range of public art can include everything from text-based memorials (Maya Lin's *Vietnam Veterans Memorial*, 1982) and observation towers (Anish Kappor's *ArcelorMittal Orbit*, 2012) to growing a field of wheat and planting a mountain of trees (Agnes Denes's *Wheatfield – A Confrontation*, 1982 and *Tree Mountain – A Living Time Capsule*, 1996) or baking pies in a public plaza

(Anissa Mack's *Pies for a Passerby*, 2002), does one identify as a visual artist, a designer, an architect, a performance artist, an activist, a researcher or a pastry chef? The potential list of professions is extensive and, to date, we have seen all of those monikers (and many more) used to identify practitioners working within the public realm. Public art's generally recognized location beyond museums and galleries raises critical questions regarding the material framing of the practice and the contested boundaries between sited works, community arts, social practice and performance works, as well as the cultural politics of aesthetic judgement.

To embrace the full umbrella of public art possibilities within this chapter, we turn to the definition of public art developed by Cartiere and Willis (2008), which argues that public art exists outside the confines of the museum or gallery and falls within at least one of the following categories:

> Concerned with, or affecting the community or the people: *public interest*
> Maintained for or used by the community: *public place*
> In a place accessible or visible to the public: *in public*
> Paid for by the public: *publicly funded*
>
> (Cartiere and Willis 2008: 9)

There has been an unspoken consensus in the fine-art establishment that 'public art' is synonymous with compromise and dilution. However, this working definition allows us to embrace the term *public art* as the signifier of a legitimate discipline, a layered practice, an area of study and a significant contribution to a broader fine art canon (Cartiere and Willis 2008: 15). Given the depth and breadth of the practice, public art as understood from this perspective can incorporate a single sited object or envelop an entire streetscape. This definition also suggests how collaboration between art and social science can have far-reaching consequences for multiple disciplines.

One of the disciplines that public art has had significant impact on is urban planning, particularly within the continuum of urban regeneration policies and revitalization schemes. The regeneration of urban landscapes employing public art opportunities could easily be conceived as restoring local memories of place or reinitiating a shared sense of civic pride. Yet the refreshed image of place generated through 'conspicuous construction' (Owen 2002) could also be accused of masking profound social and cultural crises. Within the urban context, public art has also been criticized for promoting an all-encompassing framework that constructs a space that is both depthless and value-free (Miles 1997). Therefore, when attempting to create for everyone, the diluted result is often a creation that no one responds to. Since one cannot reasonably speak of a universal public with any significant understanding towards a unique audience, the question of which 'public' public art is created for comes into question, but is still seldom analysed from a perspective other than a position of art criticism. Newbery (1999) quotes Jacobs's claim (1995: 53) that: 'Evaluation tools are not regularly employed, so the only gauge of a work's success or failure in reaching its audience is when debate erupts into protest.' This position still held true in 2015 as evidenced by the backlash aimed at public art works such as Calgary's *Travelling Light* by the artist group Inges Idee. Created as part of Calgary's per cent for art program that mandates one per cent of each city project over $1 million is budgeted for public art, *Travelling Light* is a 17 metre-tall, bright blue steel ring with two replica lamp-posts emerging from its apex. The piece was installed in the middle of an intersection between two major thoroughfares connecting industrial parks, the airport and Calgary's expansive suburbs. Unveiled in October 2013 as part of a substantial transportation department infrastructure project, the final budget for the artwork reached $470,000 and it was that figure that sparked the public debate. The public controversy seems to have begun with an opinion piece by columnist Michael Platt in the 7 October 2013 edition of the *Calgary Sun*. With the incendiary headline 'City of Calgary spends $470,000 on blue hole on far-flung stretch of road in the name of public art', Platt begins his article with the following:

> To bicker over $4,800 in suburban development costs, when the city has $470,000 to blow on bad art in ridiculous places. It seems all too fitting that Calgary's latest public art project is a giant hole – it should serve as a symbol of money thrown away, for all the good it does us. Not that there's anything wrong with public art, when it's done right, for the right reasons.
>
> (Platt 2013)

This opening quote is revealing in that it speaks to several political and social issues that were under debate in Calgary at the time that had nothing to do with this specific work of public art. The city was in the midst of its mayoral election and a $4,800 per-house levy proposal for new suburban development was being hotly contested.

> When Naheed Nenshi's entire re-election campaign based on a $4,800 per-house levy he wants to charge for new development in the deep suburbs, you'd think blowing 98 houses worth of cash on bad art would have the mayor and his aldermanic allies livid.
>
> (Platt 2013)

Caught between political factions engaged in typical election-time rhetoric, *Travelling Light*, dubbed 'The Giant Blue Ring', became the target for both politicians and the citizenry to express their dissatisfaction with issues ranging from taxation to housing. The uproar caused the city council to order a review of infrastructure funding for public art in 2014 and as Calgary experienced an economic downturn in 2015, Councillor Peter Demong put forward a motion to suspend public art funding for a year (the motion was defeated nine votes to five). The debate over *Travelling Light* and the subsequent review of public art spending did not focus on the potential social, cultural, economic benefits of public art in an urban environment; but rather highlighted the $47,000 cost of the sculpture (though not the $47,000,000 spent on the transportation infrastructure project) and subjective views of the artwork, such as the comment by Mayor Naheed Nenshi during his re-election campaign that the work was 'awful'. Post-election, the mayor later defended infrastructure projects and the public art that goes with them during the 2015 council debate about whether to suspend the public art budget.

> 'If there are additional things that we get as a result of a slowing economy, or it's time to catch our breath, we will grab them,' said Mayor Naheed Nenshi. 'But that doesn't mean that we shouldn't be moving full steam ahead on infrastructure projects, because we are way behind. So we should be using this time to catch up.'
>
> (Ramsay and Vaessen 2015)

The lack of formal evaluation tools for public art is not just a Canadian issue. One of the first comprehensive volumes on contemporary public art in Europe – *Public Art: A Reader* (Matzner 2004) – contained no references to evaluation processes, be they cultural or social. In 2002, United States art historian Harriet Senie (2002: 186) stated, 'I know of no mechanism currently in place and certainly no budget for gathering responses to a work of public art.' As of 2015, such a mechanism has still not appeared. Some have turned to social media as a means of expanding the public conversation. Another controversial Canadian public artwork (the controversy focusing primarily on its $100,000 price tag), the Main Street Poodle in Vancouver, has its own Twitter account with the profile, 'I'm 7 feet tall, porcelain, and looking for love at Main and 18th. #$100,000mutt' (https://twitter.com/MainStPoodle). A Twitter account was also set up for *Travelling Light*, by an anonymous supporter of the artwork. The profile for @GiantBlueRing reads, 'Much Maligned, Misunderstood, Majestic. A True Blue Underdog Story. Landmark Public Art Bringing People Together Since 2013. I Get Around . . .' (https://twitter.com/GiantBlueRing). However, such social media outlets serve more as parody accounts and do not engage substantive discussion or informative evaluation.

Not only are the evaluation tools still lacking for public art, but there is still inconsistency when it comes to defining or understanding who 'the public' for public art might be. In a volume on public art, McLellan acknowledges (2003: xiv) that 'the public on the street is more colourful and diverse than it is in a museum', but he does little to further investigate who that constituency might be. Debates over the nature of 'publics' in the West have a long history in the social sciences. Habermas (1989) has famously traced the relationship between publics and the emergence of post-Enlightenment, 'modern' subjectivities through participation in spaces of urban debate and interaction.

In search of a research methodology

Given the multiplicity of the public realm and the diverse range of artworks that constitute public art, responding to such questions would require a large-scale comparative project which: equates different models of development of public projects; moves beyond 'productionist' and 'semiotic' models of assessing art

(Hall and Robertson 2001), examining both proposed intentions and unanticipated outcomes of community use and sociality interpretation; and assesses responses to projects once they are in place, observing the engagements of publics forming around as well as flowing through the works. These were some of the formidable challenges that the research project chART: Public Art Marpole had to consider in order to establish a viable framework for both quantitative and qualitative analysis of the sustainable social, cultural, environmental and/or economic influences of public practice. chART: Public Art Marpole was a comparative research project that focused on how urban public art in the city is constructed, performed and consumed both visually and materially by various 'publics' (Coleman 2001).[1] One focus of the research was to examine how urban imaginaries are (or are not) embodied through spaces designed for public accessibility and collective use. Vickery's (2011) concept of 'Public Authorship' was useful in characterizing the relationship between the artists, commissioning bodies and intended publics – the three main constituencies engaged in the development, creation and reception of public artworks.

A deciding factor in utilizing Marpole as a case study for this research project was that this Vancouver neighbourhood represented a community of significant cultural diversity with a range of economic, social and environmental challenges. These challenges included significant immigration activity; a broad multicultural constituency; increasing intergenerational issues; environmental concerns regarding the delta waterway; a neighbouring First Nations community with a desire for improved relations; and a main business thoroughfare that made Marpole a place people drive *through*, not *to*. The local business improvement agency was actively looking at public art as a means of addressing some of these issues. The research project attempted to function across the community on different scales, with different social, ethnic and economic make-ups, and within specific stakeholder groups, taking into account qualitative and quantitative impact across different zones of use.

For decades, much of the research on public art evaluation has been based in the United Kingdom, yet Sarah Selwood's landmark report for the Policy Studies Institute, 'The Benefits of Public Art' (1995), stated that none of the case studies highlighted within the report had previously been fully evaluated and that the criteria by which they might have been were seldom identified. The 1999 report 'Evaluation of public art projects funded under the lottery', relied largely on telephone questionnaires. Matarasso's (1997) assessment of social impact of participatory arts programmes also drew mainly on questionnaires. Even the report by the UK public art think tank Ixia (2011) on the state of the field relied primarily on questionnaires and presented a very limited perspective related largely to responses toward the economic downturn. Little has changed between 1995 and 2015 in this regard. The chART: Public Art Marpole study therefore attempted to deploy a combination of research methods in order to gain a multi-dimensional assessment of how specific public art projects within the community relate to a wider spectrum of effects. It was necessary to tailor our research methods to bureaucracies (e.g. questionnaires, interviews, archives); to local communities and residents (e.g. direct community engagement with artists, appropriate participant observation, elicitation techniques, focus groups, interviews); and to Marpole visitors (e.g. questionnaires, observation, interviews, researcher-led video material).

As the project began, the research team proposed the following methodologies:

1 'Formative' and 'summative' evaluation, looking at process as well as outcomes; investigating attitudes *before* and *after* a project has been completed.
2 Investigating the potentially conflicting views of stakeholders, producers and consumers to discover how varied publics conceptualize criteria for 'successful' public artworks.
3 Locating the attitudes expressed in relation to wider understandings of public art, environment, identity and community using methods such as timelines and mapping techniques.
4 Combining quantitative and qualitative methods to juxtapose results of questionnaire research with semi-structured interviews and focus groups.
5 Deploying participant-observation to observe social engagements with and at sites.
6 Drawing on ethnographic, participatory research techniques following participants along their routine trajectories through the greater urban space of Vancouver, thus locating public artworks in the context of wider experiences of the city during commuting or travel for leisure.
7 Locating fieldwork within a wider economic and cultural assessment of the impact of a given artwork (Cartiere 2012).

In order to benefit from such a full range of methodologies, rather than relying primarily on questionnaires that responded to works post completion, the research project deployed a different project development approach and deployed a series of temporary public art projects in Marpole as test cases, attempting to track the artworks from the earliest planning stages all the way through to completion. In essence, we attempted to create public art as a methodology for evaluating public art.

chART Public Art Marpole: a story with small beginnings

We return to our group of artists sweating in the sun on Mother's Day, 2014, as they celebrate the completion of their street mural after two long days of kneeling on the hot pavement. In reality, their work began two years earlier on a similarly hot and sunny day at Marpole Summerfest, 2012. On that day, illustration students from Emily Carr University produced 20 small chalk street murals, while 250 community members voted on preferred styles, uses of colour, subject matter and overall favourite mural.

This event took place at the beginning stages of chART Public Art Marpole's development, when the project was not yet named as such. In partnership with the Marpole Business Improvement Association (BIA), Cartiere and Guindon successfully applied for a Mitacs Accelerate Internship Grant, which would allow the hiring of Guindon as a Masters student intern and provide $5,000 for project funding. Over the summer of 2012, the team produced five projects that, like the chalk mural contest, were temporary in nature. These works were designed to provide opportunities for the team to learn more about the community, meet stakeholders and forge partnerships. The methodology was to create the relationships necessary to make future projects happen – in other words, to use one form of public art to create other forms of public art.

For our first project, it was necessary to bring in an experienced artist who could begin our journey with an impactful work. Anthony Schrag's work, *Bench'd*, consisted of a two-day event on the street benches that ran along Marpole's main business artery of Granville Street. On the first day, Schrag staged a peaceful 'bench protest' by asking pedestrians to share something they were passionate about, writing it on a hand-made cardboard sign, and offering to protest on their behalf. On the second day, local historians, community stakeholders, city representatives, as well as a Musqueam storyteller, were scheduled and available for 'bench conversations' with participants. By the next day, no physical trace of *Bench'd* remained on site, but the connections that were made would lead to future partnerships.

The remainder of the summer projects were produced by Emily Carr students, and each one reinforced important connections with the community. From local businesses to neighbourhood houses, Vancouver transit authorities to individual residents, the team was steadily learning what the community was comprised of, what their values and ideals were, and what they envisioned for their future. Through a storytelling activity for kids (*Home*, Zara Contractor), a drawing intervention at a local coffee shop (*Show Me How*, Meaghen Buckley and Ashley Guindon), and the aforementioned chalk mural contest, the team was building a reputation and developing important contacts for their future projects.

Perhaps the most impactful of those first projects was a temporary vinyl wrap for a derelict bus shelter, *Cloud Coordinate*, an idea conceptualized by Felicia Batzloff Gail and Elisa Yon. From the beginning of our discussions, the Marpole BIA had identified a particularly problematic bus shelter as a potential site for intervention. The heavily graffitied, scratched and dirty shelter was slated for removal and had been taken off the maintenance schedule, yet still languished many months later as an eyesore in the community. After countless negotiations with transit authorities, the team received permission to wrap the shelter with an artist's design, with the understanding that the work would be removed along with the structure within the year. Batzloff-Gail and Yon developed a work that involved community engagement and photography. During Summerfest, they engaged with the community from within the bus shelter, handing out postcards of their site photography (which captured the streetcar lines set against a cloudy blue sky). On the postcards, participants answered the question: 'If you could go anywhere in the whole wide world, where would you go?' The responses were collected into an artist's book (which was later placed in the Marpole Library), and the images from the postcards were applied to the shelter, creating a translucent, ethereal wrap. The response from the community was positive, with many expressing their loss when the work was removed as expected. While *Cloud Coordinate* did not last for more than a year, the feedback from the project further cemented our reputation and track record in the community. It would also allow us to return in 2014 to the same partners and wrap not one, but *seven* bus shelters along Marpole's main business artery. Artist Nathalie Lavoie's photographic series *Meniscus* remains in Marpole still, which would not have been possible if not for its predecessor.

In the fall of 2012, the Mitacs grant was exhausted, and Cartiere was able to acquire a Social Sciences and Humanities Research Council (SSHRC) Insight Development Grant with the aim of producing five further projects over two years. At this point, it became necessary to name the project. With so many partners and no physical location in the community, the team found that there was confusion as to who originated the works: were they produced by the City of Vancouver, Marpole BIA, Emily Carr University or Translink? An identity was therefore necessary. A community engagement workshop at the local Neighbourhood House (led by Emily Carr communications experts) would result in the project's name, chART: Public Art Marpole. Invoking themes of searching, mapping and connection, this identifier would help to unify the projects regardless of their style, theme, partnerships or location. A photography project at a local high school, seasonal street banner or interactive new media work could fall under the same umbrella. chART as a name became the common denominator for everything and eventually became known as a community entity in its own right.

Stepping stones: using public art to create more public art

With new funding and a strong identity in place, the team was ready to build on the initial connections established with the first five Mitacs projects. The 2012 chalk mural contest mentioned at the beginning of this chapter evolved into several related works of a semi-permanent nature. At that first Summerfest, the team had collected 250 ballots (a 100 per cent return) in which participants shared their artistic preferences. The team noted that visitors who viewed the chalk murals without a ballot spent on average 1.5 minutes viewing the work. Visitors with a ballot spent an average of 5.5 minutes viewing the work. These ballot-carrying visitors were also more actively engaged in conversation with other visitors about what they were seeing, how and why they were voting for particular works, and more likely to talk with the artists who were still drawing away. The information garnered from the ballots (preference for style, colour use, themes, etc.) was given to an illustration class at Emily Carr as guidance for the semester-long development of four proposed street mural designs. A jury (comprised of Marpole community members and Emily Carr faculty) chose two of the designs. Six months later, and almost a year after the chalk mural contest, the first street mural – *Fractal* (Figure 5.1) – was completed (despite delays due to Vancouver's unreliable weather). The second street mural – *Leaves and Fishes* – took another year to complete for the same reasons. In addition to the murals, the team was able to work with the City of Vancouver's Graffiti Management Program to

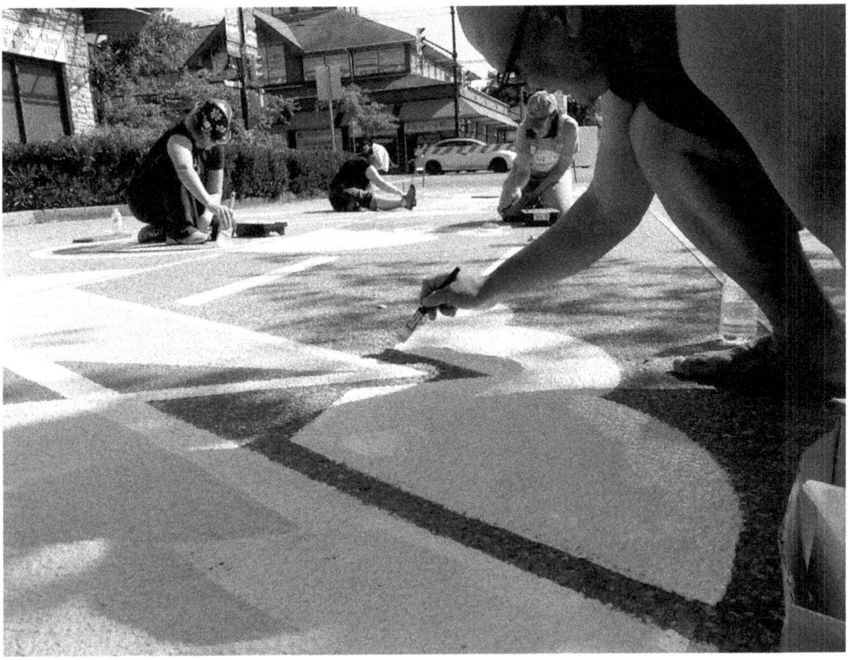

Figure 5.1 *Fractal* by Alejandro Quinteros (mural in progress), 2013, Vancouver. Source: Ashley Guindon.

explore utility wraps as vibrant and effective public artworks. The winning chalk mural design was turned into Marpole's first electrical box wrap, and two other student-designed wraps serve as mural didactics (one electrical box wrap, as well as Vancouver's first Hydro pole wrap).

By building on smaller, temporary and lower-risk projects, our team was able to improve the work's quality and quantity, thus increasing overall impact over time. Indeed, the five projects Cartiere initially proposed for the SSHRC grant turned into 22 projects by the time the grant finally ran its course. Many of these works were still temporary or small-scale (such as two artist-led classes at a local high school, more utility box wraps, and a set of street banners produced through a contest for Emily Carr students; see Figure 5.2) and continued the important work of building relationships, creating opportunities and building chART's brand. However, other projects took on more ambitious challenges and had broader impacts. *Park-a-Park*, a mobile park built from a roll-off dumpster, was conceived by designer Julien Thomas; by combining the creator's aims with the mandates of our SSHRC grant, our team was able to fund the project and facilitate its development. Designed to fit in a standard parking space, the mobile park was hosted at businesses, schools, and real-estate developments in Marpole and around Vancouver. Similarly, the team identified Sans Facon's

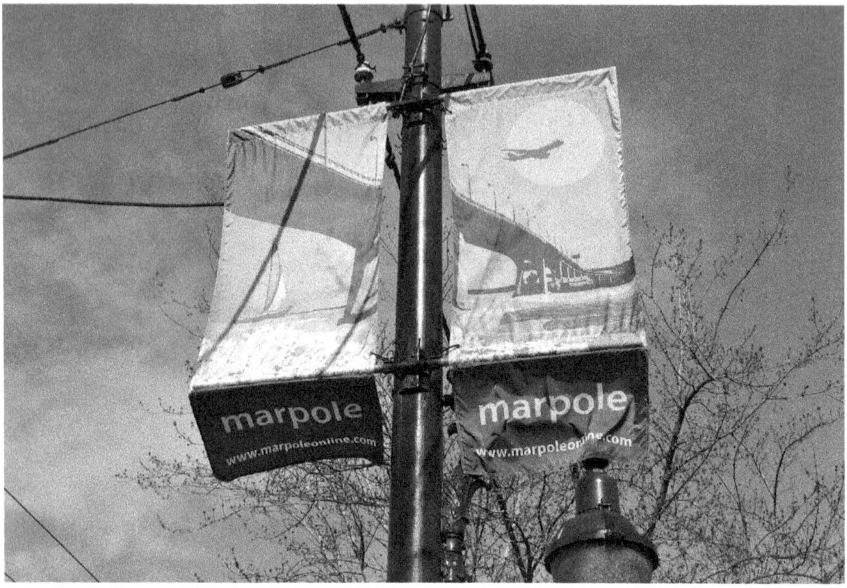

Figure 5.2 *Untitled* by Set Narathipphatthara (Marpole BIA banners), 2014, Vancouver. Source: Cameron Cartiere.

project *Limelight: Saturday Night* as an innovative event that would contribute to our work in the city. The project is a street intervention in which two street lamps are replaced by spotlights, creating a limelight on the sidewalk for passersby to interact with, and the creators have showcased it globally. By liaising with the Marpole and Downtown BIA's and securing a Canada Council for the Arts Grant for extra funding, we were able to produce *Limelight*'s Vancouver edition – the first time the project occurred in two locations at once. These projects and others like them went a long way toward increasing chART's visibility and reinforcing our methodology.

Recognition, however, is not the only marker of an effective project. Perhaps our most successful work was not an original idea, nor was it large, expensive or high-profile. As part of our growing partnerships with community stakeholders, we had developed a relationship with Marpole Neighbourhood House – a community hub and the home of many local organizations. Our plan was to support a student to design a Free Little Library (one of many in the international movement) for the House. The plan was accepted, the design was completed (adorned with a bell tower to match the House's most distinguishing feature: an old firehouse tower), and the library was built. Then, suddenly, the House experienced a destructive flood that shut down all operations and dispersed the organizations. The majority moved down the street to a local church, and this temporary home soon became permanent as municipal challenges kept the original site closed.

In the end, it was decided that the library would be reinstalled at the church; as luck would have it, the design was still relevant as the firehouse tower became a church steeple. Within six months, this temporary solution had transformed an empty patch of grass alongside the church into a vibrant public space. Planter boxes, bird feeders and benches were later added by the community, who unified all the additions with a matching coat of paint. The library itself was granted an award of distinction and featured on the Free Little Library website. Perhaps most importantly, the library is used daily by the neighbourhood and there is a continual turnover of books being shared by the community. 'Ownership' of the project has transferred completely from chART to the neighbourhood, not by an official transaction, but by community use and connection to the site.

Looking back: the 20/20 lens of hindsight

What makes a work a success varies greatly depending on the critic, but the fact remains that five projects turned into 22, and the project continues with great community support today. Perhaps this can be attributed to the long-term nature of the work. It is common practice for an artist to come into a community, produce a work and then depart – a phenomenon sometimes called 'parachuting'. With this model, the community is often left with a creative gap or loss after the artist's departure. Moreover, the next artist to come along must start again at the beginning: forging relationships, building trust and proving his or her social credibility. Producers have responded to these issues of temporality in other communities using a similar 'branding' strategy as we used with chART. Particularly noteworthy is the work of Situations, based in Bristol. Similar to chART, the Situations project started as a research initiative with the University of the West of England (UWE). Led by researcher and writer Clare Doherty, the project evolved over time to become its own organization operating independently from the university. Rooted firmly in the city of Bristol, Situations has been able to build a reputation for innovative work which over time has become more and more experimental, while still maintaining the support and enthusiasm of the local community. Situations and other organizations (Platform, Common Ground, Project Row Houses) are examples of a growing number of researchers and creative practitioners who have identified the value of a long-term investment in one locale.

In the case of chART, the core team was the anchor for the overall project. Our relationships with stakeholders carried over from piece to piece, so that incoming practitioners could tap into the network we had already established. Moreover, because we had garnered trust in the community, the team could afford to take more risks. In particular, we were able to create opportunities for emerging artists who were just beginning to consider public art as a potential field. For example, one of the co-creators of *Cloud Coordinate* had never done public art before; now, she is a public art project coordinator for a local municipality and has continued to produce work in the public realm. Because of the team's constancy and consistency in the community, first opportunities – so important for an emerging career – were more readily available. Artists not yet out of school were provided with a lived experience and an opportunity to place real-world examples of their creative talents in the public realm; in fact, 11 of the 22 projects produced since 2012 were created by graduate and undergraduate students.

Perhaps the most positive outcome of chART's approach is the fact that projects that would normally be considered failures were, in fact, just as useful to the team's aims as those that were seen as successful. For example, two projects proposed for the Marpole Oakridge Community Centre (MOCC) – a community hub and an integral partner identified by the team since the beginning – did not come to fruition. One was *Park-a-Park*, which was slated to reside at MOCC temporarily. Unfortunately, insurance issues prevented its arrival at the last minute, so the project was relocated elsewhere. The second project was a series of workshops at the community centre led by a local artist. Despite hours of meetings and negotiations, the project did not receive its final approval and never materialized.

When a project falls through it is always disappointing, even though it is all-too common when working in the public realm, where mandates change, priorities shift and interest waxes and wanes. Unfortunately, the loss of such projects often means the end of the work in that community. However, because of our long-term approach, these losses had the time to be turned back into new opportunities. In the case of our relationship with the MOCC, we were able to demonstrate perseverance in the neighbourhood and stay on the organization's radar. In time, the organization came back to us with a more in-depth opportunity – a long-term commitment to be the artists in residence at the MOCC's Fieldhouse. The fieldhouse scheme is managed by the Vancouver Parks Department and places artists into small houses sited within the parks system, with the mandate of community engagement through the arts. Many of these houses are situated near community centres and when the fieldhouse opportunity arose in Marpole, the MOCC recognized that chART was the perfect fit. Perhaps the surrounding neighbourhood, so used to seeing the house empty and derelict, perceives our arrival as sudden, albeit welcome. But our physical presence in fact began years previously with these unfulfilled community centre proposals – the conversation starters that opened the door for a continued partnership.

It is also interesting to note how our relationship to the community has changed over time. This is perhaps most evident in the way we have participated in the community's challenges and trials. In 2013, a controversial new Community Plan was developed to replace Marpole's 30-year-old version and the response from the community was immediate. Stakeholders protested at city workshops, and 'Stop Marpole Rezoning' signs covered numerous lawns. Though we were invited to the city's workshops, we ourselves did not wade into this debate. We did, however, propose and facilitate our own workshop for our partners and stakeholders. In collaboration with the BIA and a local credit union, we brought in a strategic expert to run two events – one for 'insiders' (Marpole stakeholders) and one for 'outsiders' (representatives of organizations from the greater Vancouver area with an interest in Marpole). The findings of the workshops, which questioned where Marpole came from, where it was at present, and where it could go, will lead us into the future of chART. For now it is enough that we were trusted with the facilitation of this workshop and tasked with the development of solutions.

At the beginning of the project, we were sometimes met with apprehension from Marpole residents. 'But do you come from Marpole?' was a common question, as well as 'Why are you doing this here?' Our answer was always 'We were invited', as indeed we were from that first email sent at random to Emily Carr University from the Marpole BIA that ended up in Cartiere's inbox, to each project in which we were invited back by the partners, stakeholders and community at large. Now, with the fieldhouse as a home base in the community and four years of engagement behind us, it is perhaps safe to say that we do come *from* Marpole.

Looking backwards and forwards for conclusions: methods of evaluation

Public art often suffers from immediate (and often hasty) media and community reactions to change. A new work arrives and the spotlight of criticism is focused on it – for about a week. Unless it becomes the centre of some civic controversy like *Travelling Light*, the work will slip from the public consciousness with the next news cycle. It is after that immediate reaction, that the real impact of the work begins to filter into the urban fabric. If the work is a permanent feature, the evaluative questions often revolve around its use and maintenance. Do people make specific trips to see the work? Has the work been vandalized, and if so, how often? Has the work improved the quality of the streetscape? If the work is temporary, the evaluative questions revolve more around the myth and memory of the work (Lacy 1995). Does the community remember the work a year later? Did the temporary work result in other works within the community? Has the documentation of the work filtered into the history of the community? With a long-term research project such as chART: Public Art Marpole, when has enough time passed to begin substantive evaluation?

As we move forward with the research, the initial aims of the project are starting to become achievable. With the goal of obtaining quantifiable data in mind, and with the overwhelming desire from Marpole community members for environmentally sustainable initiatives at heart, Cartiere has developed the *Border Free Bees* project. This project is a collaborative revitalization initiative to raise awareness of the plight of wild pollinators, empower communities to actively engage in solutions for habitat loss, and transform under-utilized urban sites into aesthetically pleasing and environmentally sustainable spaces that serve as pollinator pastures as well as public spaces. The project is sited in the cities of Richmond and Kelowna and managed in partnership with Emily Carr University, the City of Richmond, University of British Columbia Okanagan and the City of Kelowna. This project initially began as a proposal developed for Marpole (that has not yet manifest), and resulted in a substantial project across the Lower Mainland. We foresee that all the work that is happening outside of Marpole will loop back once the opportunity arises to bring the project back into the community.

The tracking and feedback methodology that is being developed for *Border Free Bees* is a direct result of our experiences in Marpole, which allowed us to utilize, experience and evaluate several methods of tracking and identify those that were most effective. Traditionally, methods of evaluation are considered an extra step at the end of a project, but we have found that when they can be used as an integral part of the work, rather than separately from the work, an evaluation method is generally more engaging for people to participate in (and its probability of capturing actionable results increases accordingly).

For example, our Summerfest 2012 Chalk Mural Contest generated a high return on our method of evaluation because we empowered people right from the beginning by telling them that their opinions mattered and would have an actual result by not only affecting the artist who won the contest but also later by inspiring a new community artwork. While one would argue that the success of the mural (and the fact that it has not been tagged or vandalized) is a result of it meeting the graffiti management requirements, it could also be argued that a community connection has generated an additional impact. In this case, then, the method of evaluation not only provided feedback, but it also boosted the project's success. With this in mind, the next phase of the research includes evaluation methodology as part of the public art planning and a built-in element of the public artwork.

It is also clear to the team that chART: Public Art's impact carries on beyond what we can tangibly observe and measure. For instance, the team took a chance with a small vinyl company for their first electrical box wrap – a commission the firm had not encountered previously – and we have gone back to them for all similar projects, some of which were very technically challenging indeed. The initial impact of this interaction is clear, as not only has the company continued to benefit from our business, but they have also had to stretch beyond their typical capabilities and develop new technologies, as when the company took on the challenge of helping us to wrap water barrels for a community garden, leading to the production of the first light-post wrap in Vancouver. Such initial impacts are easily measurable, but their ripple effects are complex and harder to evaluate. For instance, the company has now been retained by the City of Vancouver to produce electrical box wraps outside of our project. While it is possible that the city might have discovered this company if they had not been involved in our project, our working together certainly helped put them on the map faster, creating new opportunities for their business. This type of economic impact is important data that can greatly support municipalities as they attempt to justify public art expenditure. As *Border Free Bees* progresses, these and other impacts can be tracked with the help of our partnerships with other institutions and municipalities.

With *Border Free Bees*, Cartiere's research focus shifts to environmental sustainability on a scale beyond that of a single community and the question of the chART's project status in Marpole arises, along with greater questions about the long-term nature of public art. When does a project run its course? Is artist or community fatigue a factor? Is it better to prepare for one's own obsolescence? Or is there room for the exponential expansion of creative energy?

Because of Marpole's reputation as a community in transition, our experience indicates that there is still much potential for public art initiatives in Marpole. As we move forward with our new initiatives, we retain our commitment to community engagement as a cornerstone of the work. Too often, public art can be the outcome of a development project, an unsolicited donation or the re-siting of a work from another location, with little significance or relationship to that site. As a result, this type of 'plop art' can feel like a cast-off that sticks out like a sore thumb, making the community feel that their area is treated like a dumping ground for unrelated projects. Herein lies the distinction between historical sited works, 'plop art', site-specific work and community-engaged practice – all of which we have grappled with in Marpole. For our purposes, we will continue to argue that one cannot create a site-specific work without taking the time to understand that site (its history, people, and so on): one cannot produce effective social practice without an engaged audience.

Another key element of our work is our attention to the ongoing care and maintenance of a public art piece. This aspect of the work is harder to sell than its initial installation – too often, decision makers are ready to move on to the next thing once the initial piece has been commissioned. If we want the work to be impactful for the long term, then its full life-cycle has to be considered and integrated into its planning from the beginning. The urban environment is constantly changing, and a work left unattended will eventually become aged, vandalized or obsolete. Taking a sustainable view, rather than a short-term one, may not be popular, but a cradle-to-grave approach (McDonough and Braungart 2010) should be part of the plan from the beginning, as well as planning for the work to gracefully fade (after all, community street murals do not generally last more than one or two years). This is why working with partners who have a long-term investment in the community is so important, as the team can integrate the idea of tending to the full life-cycle of a work into the partner's plans from the beginning. Even if the work is open-ended, one must still consider how that open-endedness is managed over time.

With these core principles in place, our work continues as some earlier initiatives have finally come to fruition after years of relationship building. The Marpole Oakridge Community Centre and Vancouver Parks Board have awarded chART residency in the Oak Park Fieldhouse (the latest addition of field houses to the Vancouver Park's Department's ongoing artist studio program). With Emily Carr University as a partner, the Fieldhouse will serve as a hub of public art activities, allowing students from Emily Carr to come for Marpole-focused classes while visiting artists, environmentalists, writers and designers are already lined up to produce community events.

Another initiative involves initiating more complex community partnerships and pursuing funding strategies for future projects in the community. As part of her professional practice, Guindon has developed a wall mural, with accompanying community engagement, for the Marpole Library's exterior walls. The project is a partnership between chART: Public Art Marpole, the Vancouver Public Library, the Marpole BIA and Joy Kogawa House (a local writer's residence and the historical home of Canadian author Joy Kogawa). This is the first of our projects produced outside of the academic research grant, and exemplifies a continuation of the past four years' efforts. This work also marks a shift in the sustainability of chART as a working entity. As Cartiere's focus moves more toward projects outside of the neighbourhood, Guindon has been able to transition from the role of research assistant to project manager for the Marpole-based work.

These expanded opportunities are made possible because of the relationships we have built and also because we have made a commitment to the continuation of the work. By accepting the responsibility of the fieldhouse, we now have a physical anchor in the community as well as a deeper personal connection to the place. The opportunity could not have happened without the web of community relationships built over the past four years. But like any web, such relationships, even developed over a long period, remain fragile and easily broken. Sustainability means not only a focus on the development of the work, but the ongoing maintenance of the community relationships and the evolution of the practice that brings the work to fruition. Each season requires a renewal of purpose and a rededication to the collective project.

Note

1 The inspiration for this research stems from a research proposal developed several years earlier in collaboration with anthropologist Simon Coleman (Chancellor Jackman Chaired Professor in the Department for the Study of Religion at the University of Toronto), exploring the concept of 'Spectacular Cities'. This original proposal developed from the position that cities in post-war Britain have become increasingly aestheticized and that older models of urban structure have been challenged by new, spectacular, 'ludic' forms, such as malls and waterfront developments. As a result of this process, public art has been regarded as a key resource in the regeneration of such urban areas.

References

Cartiere, C. 2012. Charting public art – a quantitative and qualitative approach to understanding sustainable social influences of art in the public realm. *Art & the Public Sphere*, 2(1–3), 9–16.

Cartiere, C. and Willis, S. (eds) 2008. *The Practice of Public Art*. New York: Routledge.

City of Calgary. 2004. Corporate public art policy [Online: Calgary.ca]. Available at: www.calgary.ca/CSPS/Recreation/Documents/Public-art/Public-Art-policy.pdf [accessed: 28 June 2015].

Coleman, S. 2001. Grounded tourists, travelling theory, in *Tourism: Between Place and Performance*, edited by S. Coleman and M. Crang. Oxford: Berghahn, 1–17.

Habermas, J. 1989. *The Structural Transformation of the Public Sphere: An Inquiry into a Category of Bourgeois Society*. Cambridge, MA: MIT Press.

Hall, T. and Robertson, I. 2001. Public art and urban regeneration: advocacy, claims and critical debates. *Landscape Research*, 26(1), 5–26.

Jacobs, M. (ed) 1995. *Sculpture Chicago: Culture in Action*. Seattle, WA: Bay Press.

Lacy, S. (ed) 1995. *Mapping the Terrain: New Genre Public Art*. Seattle, WA: Bay Press.

McDonough, W. and Braungart, M. 2010. *Cradle to Cradle: Remaking the Way We Make Things*. New York: MacMillan.

McLellan, A. (ed) 2003. *Art and Its Publics: Museum Studies at the Millennium*. Oxford: Blackwell.

Matarasso, F. 1997. Use or ornament? The social impact of participation in the arts. Report for Comedia, London.

Matzner, F. (ed) 2004. *Public Art: A Reader*. Ostfildern-Ruit: Hatje Cantz.

Miles, M. 1997. *Art, Space and the City: Public Art and Urban Futures*. London: Routledge.

Newbery, C., Britain, I. and Hicter, M. 1999. Evaluation strategies for landmark art projects based on a comparison of three European projects. *Evaluation*, 8(9), 2–52.

Owen, K. 2002. The Sydney olympics and urban entrepreneurialism. *Australian Geographical Studies*, 40(3), 323–336.

Platt, M. 2013. City of Calgary spends $470,000 on giant blue hole on far-flung stretch of road in the name of public art [Online: *Calgary Sun*]. Available at: www.calgarysun.com/2013/10/07/city-of-calgary-spends-470000-on-giant-blue-hole-on-far-flung-stretch-of-road-in-the-name-of-public-art [accessed: 29 July 2015].

Ramsay, M. and Vaessen, D. 2015. Public art funding to remain unchanged [Online: Global News]. Available at: http://globalnews.ca/news/1845250/city-councillors-vote-against-suspension-of-2015-arts-funding/ [accessed: 29 July 2015].

Selwood, S. 1995. *The Benefits of Public Art: The Polemics of Permanent Art in Public Places*. London: Policy Studies Institute.

Senie, H. 2002. Responsible criticism: Evaluating public art. *Sculpture*, 22(10), 186–190.

Vickery, J. 2007. The emergence of culture-led regeneration: a policy concept and its discontents [Online: Working Paper, University of Warwick, Centre for Cultural Policy Studies, Warwick, United Kingdom]. Available at: http://wrap.warwick.ac.uk/36991/ [accessed 15 May 2015].

Vickery, J. 2011. Art, public authorship and the possibility of re-democratization. *Visual Culture in Britain*, 12(2), 219–235.

Zuidervaart, L. 2011. *Art in Public: Politics, Economics and a Democratic Culture*. New York: Cambridge University Press.

Part II

Affect

6 Shaping subjects, connecting communities, imagining futures?

Critically investigating *Play Your Place*

Harriet Hawkins and Ruth Catlow

Introduction

'Making is connecting', 'technologies of connection', 'politics in action' (Gauntlett 2011, McNally 2015, Toscano 2009). These are just some of the claims made for the power of contemporary art forms. That the production and consumption of art does work in the world is not a new claim of course, but such claims have come to dominate discussions of public art and participatory art within and without the academy. Indeed, whether we look to critical social theory's 'artistic turn', the growth of funding for environmental arts practices or the way governments around the world have folded arts practices into wider programs for social well-being and urban renewal, we see widespread evidence of the currency given to the 'social work' that art has been understood to do (Jackson 2011). As such, aesthetic practices are taking on a socially instrumentalized form, connecting diverse communities and fostering ethical relationships between people and their places and the environment (Kester 2004, 2011).

It is, of course, easy to get carried away on the tide of excitement around the possibilities that are opened up by arts practices understood in this way. As yet, however, as art theorists from a range of different perspectives make clear, there is a lack of understanding of the form and nature of the encounters and relations that are constituted with and through arts practices (Bishop 2012, Hawkins 2014, Kester 2004). This is especially odd given that the widely recognized social value of the arts, and their possibilities for connecting people and places, presents an imperative to reflect critically on the dynamics and value of these relationships. As a range of key art theorists have noted, there is a need both to move beyond taking such connections and relations at face value and to ask deeper questions regarding the form and kind of these connections. This is to query, in short, the forms of relationship that arts practices can produce between individuals, and with respect to the individuals and the communities and environments in which they live (Hawkins et al. 2015).

This chapter will use ethnographic work with the participatory arts project *Play Your Place* to query the nature and possibilities of the connections that arts practices might produce. *Play Your Place* is described by its creators as 'an open participatory artwork for people to shape their own towns through play, for the health and prosperity of all'.[1] Conceived and developed by artists Ruth Catlow (UK) and Mary Flanagan (USA; Flanagan 2009), it takes the form of an online and mobile game in which people come together to draw, make and play the future of their communities. It begins with a collective process of drawing place and community, and concludes with the production and playing of video games based on the challenges and future imaginaries of place that the drawings captured. In this chapter we bring together findings produced as one of the project creators of the project (Catlow) and as a geographer who conducted 18 months of ethnography on the project (Hawkins) at iterations of *Play Your Place* at sites on the east coast of England (Southend-on-Sea) and in London (South Westminster). While *Play Your Place* has been executed in communities around the UK and Europe, in this chapter we focus in particular on its South Westminster iteration in order to query the effects it had on the participants it engaged. From such empirically focused arguments we seek to make wider points about how it is we can conceptualize the social work art does.

The argument in this chapter unfolds in three sections. The first 'Play the change you want to see', introduces *Play Your Place* in more detail and examines the critical backdrop across which our largely empirical discussions evolve. The second section, 'Shaping Subjects?', examines how the experience of participating in *Play Your Place* was shaped through practices of drawing. The third section, 'Connecting Community?', explores the process as one of mapping in order to understand how it is that we might think about the project as connecting and reconnecting existing communities and creating new ones. The chapter concludes by reflecting on the limitations of this study and the wider lessons that can be taken forward for thinking through encounters with art.

Play Your Place 1: play the change you want to see

Along the top of the opening screen of the website 'Play South Westminster' a cartoon airplane (drawn by Oliver, 36, South Westminster) pulls a hand-written banner reading, 'What will you save?' A player, jumping from one platform to the next, can choose from a series of games, each of which encourages an engagement with one of the many social challenges facing the South Westminster community. The virtual world of this city ward is one where Boris Johnson, portrayed as a Machiavellian Mayor of London, rides his community bike high above the city streets (Figure 6.1). On these same streets, caricatured figures of smartly dressed 'gentrifiers' are urged to 'grab the cash . . . be evil, kill community spirit, push up rents, hike house prices and force out families'. Elsewhere in this virtual city space, a small mouse dashes around platforms/streets armed with a fire hose trying to put out the burning buildings, the last line of defence after the closing of fire stations across London (Figure 6.1). The cute and whimsical world of South Westminster the community portrayed in their games engages us with some of the lived experiences of this city ward. For if one view of South Westminster focuses on the millionaire's mansions, leafy parks and international art galleries for which the area is famous, these games encourage us to look beyond this to see a story of

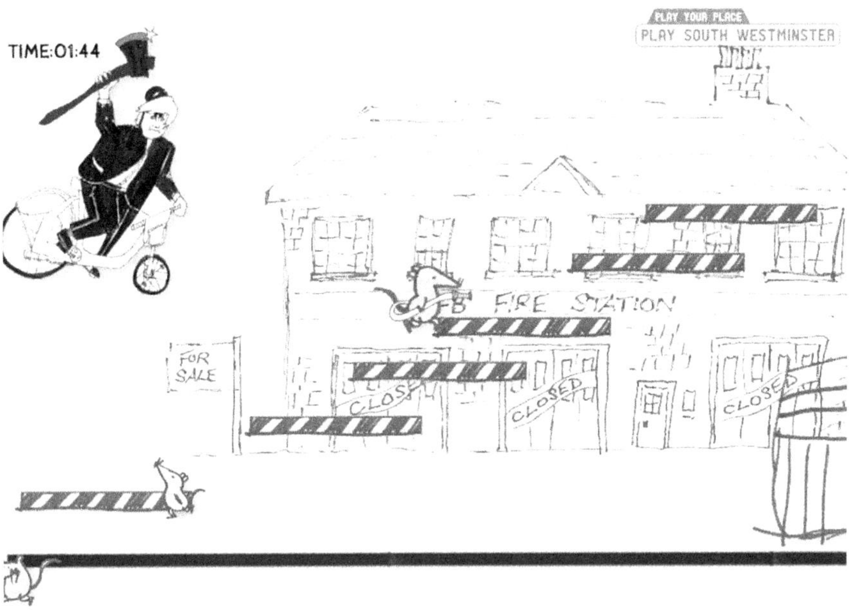

Figure 6.1 Screenshots from *Play South Westminster* game. Source: Harriet Hawkins and Ruth Catlow.

worsening inequality, a decline in local community and a loss of services under government austerity.

Play Your Place is designed as open-source platform that has toured to towns and cities around the UK and Europe, encouraging each community to draw and play the change they want to see in their place. The South Westminster iteration received additional funding from Tate Local, the community engagement arm of Tate Britain, the international art gallery based in the area. The project had four key points of intersection with the local community. The first took place at 'South West Fest', a large community-focused summer festival, during which hundreds of members of the community participated in drawing and storyboarding games about their local area. Sitting at tables, adults and children drew and painted their issues, brainstorming the community challenges with family members, friends and others in the community. Many then described their drawing and idea for a game into a camera, encouraged through discussion and prompts to articulate their ideas for community change. The second and third sites of community engagement focused on Tate Britain. Here community members were invited to participate in a week of workshops alongside professional video game designers in order to evolve the drawings and ideas into playable games. Following the workshops the finished digital games where displayed for playing during a public opening in association with the Tate Gallery's summer blockbuster exhibition *Folk Art*. The fourth site of public engagement is the ongoing presence of the games on the *Play Your Place* website, where they can be played by members of the South Westminster community as well as anyone else with a web connection and the inclination.[2] In this analysis we are less interested in exploring the finished games as visual artefacts, and rather more concerned with what happened in the course of the participatory processes of drawing, making and gaming that sit at the heart of the project.

The field of socially engaged art is fractured in its theoretical and philosophical dimensions, variously taking in art as social practice, dialogic practice, as a technology of connection and perhaps most controversially as relational aesthetics. What seems to unite these different schools of thought is a sense of these forms of art as 'affects the public sphere in a deep and meaningful way' (Helguera 2011: 7). Whether termed relational aesthetics or not, these artistic practices are understood primarily through the 'inter-human relations which they represent, produce or prompt' (Bourriaud 2002: 112). As such, the focus of aesthetic judgement in these works is around the social relations that the work constitutes rather than in the creation or judgement of their material form. In short, this is art that goes to work on the social – human and non-human – in some way, rather than merely rendering visible social issues. How exactly the public sphere and individual subjects are changed and what role art plays in doing this and how it does it, remain questions that underpin discussions of these forms of artwork.

Perhaps two of the best developed but also most controversial framings for these discussions are offered by art theorist Grant Kester (2004, 2011) and curator and theorist Nicolas Bourriaud (2002). Kester's dialogic understanding of socially engaged arts practices focuses on exploring how art catalyses spaces for conversations between its participants. Art here is understood to create situations or spaces of articulation 'where individuals can break free from pre-existing roles and obligations, reacting and interacting in new and unforeseeable ways' (2004: 6). As such, participating in artworks enables a 'challenge [to] conventional perceptions . . . and systems of knowledge' (2004: 3). For Kester, art practices make possible unconventional and challenging speech acts that 'help us speak and imagine beyond the limits of fixed identities, official discourse, and the perceived inevitability of partisan political conflict' (2004: 8). For Bourriaud, art also offers a means to develop other 'possible universes' (2002: 9). His Marxist-influenced theory of relational aesthetics explores how 'the role of artworks is no longer to form imaginary and utopian realities, but to actually be ways of living and models of action within the existing real' (2002: 13). Central to this bringing about of new ways of living is art's creation of 'specific sociabilities' constituted within 'perceptive, experimental, [and] critical' spaces of communication (Bourriaud 2002: 12–16). Such spaces operate in contrast to a generalized economic environment that reduces communication and social relations to predictable exchanges driven by commercial transactions between atomized, individually motivated subjects. In contrast, Bourriaud's relational aesthetics proposes the creation of 'social interstices'; an aesthetic rendering of Marx's understanding of alternative trading communities who operated in an economic interstice by way of other trading policies that eluded capitalist economies (2002: 16).

Our point in recounting these debates is not to try to summarize their complexities, or to present these individuals as the authorities with which we agree. It is rather to illustrate the kinds of claims made for art as offering the chance to imagine and engage new futures, to open up and enact other ways of being individually and collectively, and to offer sites and spaces of articulation and practice for disempowered groups. These are, of course, quite some claims to make about what art can do. Quite often accompanying such claims are clear articulations of the need to move away from assertions to actual evidence garnered from empirical work.

In the extended empirical discussion of *Play Your Place* which follows, ethnographic research forms the basis for an examination of two ways that *Play Your Place* might be thought to shape the social; imagining futures for places but also bringing about shifts in subjectivities and communities. The intent is not to map *Play Your Place* onto any of the art theories discussed above, but rather to stay with the lived and observed experiences of the project in order to explore what happened in one of its iterations in South Westminster. The focus of attention here are the practices of drawing and mapping that constitute *Play Your Place* at this and other locations. To foreground such processes is to shift discussion away from the content of the games, or focusing on the games as primarily representational artefacts to explore instead the effects of the participatory processes themselves.

To understand these art encounters, we are concerned with individual and collective observations and accounts of the embodied and social processes that constitute these art works. We turn away from art theory to seek other resources to understand how these practices shape the social. We find our resources here in participatory research methods and their attentiveness to what can happen in the dynamics of participatory doing. For some, to call socially engaged art 'social practice' is unhelpful; here, however, we suggest thinking through the actual lived and practiced experiences of participants in these arts forms is an important route to exploring what it is that art encounters do in the world.

Play Your Place 2: shaping subjects?

> Being the author of your own life, of the way you move through the world, is a fundamental idea in democracy. We want to help everyone use the power of storytelling to project their authority . . . getting beyond our predicaments – whether as individuals or as a society, needs to be a process of telling stories about our lives.
>
> (Lambert 2013: 2)

Writings on participatory storytelling appreciate, as the quote above demonstrates, how it is that these practices might shape the subjectivities of those involved in these creative practices. In our case, we witnessed the project engaging a wide cross-section of the local community in democratic processes, helping them to explore and assert their local knowledge and lived experiences that might otherwise be overlooked or undervalued. While 'storying' was an important part of

Play Your Place – as participants were asked to explain their drawings to a camera, creating visual and audio records of these stories – equally important to this discussion are the practices of drawing.

Key to our exploration of what the process of drawing does in *Play Your Place* is the literature on participatory visual research methods, wherein participatory research is understood as the 'accessing and valorizing [of] previously neglected knowledge and [as] providing more nuanced understandings of complex social phenomena' (Kesby 2002: 423). Within this, the claims of visual methods are multiple and oftentimes the focus for participatory researchers is on both the processes of the production of 'different' knowledge as well as the outcome of this, 'seeing people co-creating their reality through participation, through their experience, their imagination and intuition, their thinking and their action' (Reason 1998: 262). Such outcomes are nuanced further in accounts that understand participatory research to work at the level of the participating subject 'not only producing knowledge and action directly useful to a group of people' but also 'to empower people at a deeper-level through the process of constructing and using their knowledge' (Nieuwenhuys 2004). As such participants in these processes are not only empowered to speak for themselves, but are also supported in changing themselves and the worlds around them; individuals and collectives are 're-orientated, focused and energized . . . towards knowing reality in order to better transform it' (Enright and O'Sullivan 2010). We turn now to explore how this happened within the drawing processes in *Play Your Place*.

'If you draw, it will happen'

Drawing has been the subject of much art theory; rarely, however, have art theorists engaged with drawing as part of participatory arts practices, wherein it takes up a much more amateur role. More often drawing theory pivots on tensions between understandings of drawing that focus on the production of artefacts, or understandings of drawing that are focused on what happens during the drawing process (Berger 2007). Within *Play Your Place* we see both these ideas of drawing at work, as well as a sense of drawing understood from within thinking about participatory methods. Drawing can be situated within a suite of participatory visual practices (photography, collective video making) of which two are especially pertinent; first, the sense that such participatory methods are collaborative and empowering. Image work is often understood to aid in research 'with' rather than 'on' often hard-to-reach groups such as children or the mentally ill (Young and Barrett 2001). For others, what participatory visual methods potently offer is a better grasp of participants' lived experiences and practices, of how people experience their worlds. As Bagnoli notes, 'the inclusion of non-linguistic dimensions in research, which rely on other expression possibilities, may allow us to access and represent different levels of experience' (2009: 547). Second, visual methods are seen to make possible attempts 'to capture alternative vocabularies and visual grammars that are not always encountered or expressible in oral interviews' (Tolia-Kelly 2007: 135). As O'Neill expands, 'knowledge is produced forcing us to abandon instrumental rationality and reach towards a more sensuous understanding that incorporates feelings as well as cognitive reflections' (2008: 7). Such discussions have tended to focus on what participatory visual methods make possible in terms of knowledge, perhaps less clearly articulating what happens in the processes of producing such knowledge. As the following discussion of *Play Your Place* makes clear, what is interesting is not just the 'data' produced by and through the drawings, but also what happens in the process of drawing itself. These methods are, in short, less about enabling better research and rather more about transforming subjects and communities through the process.

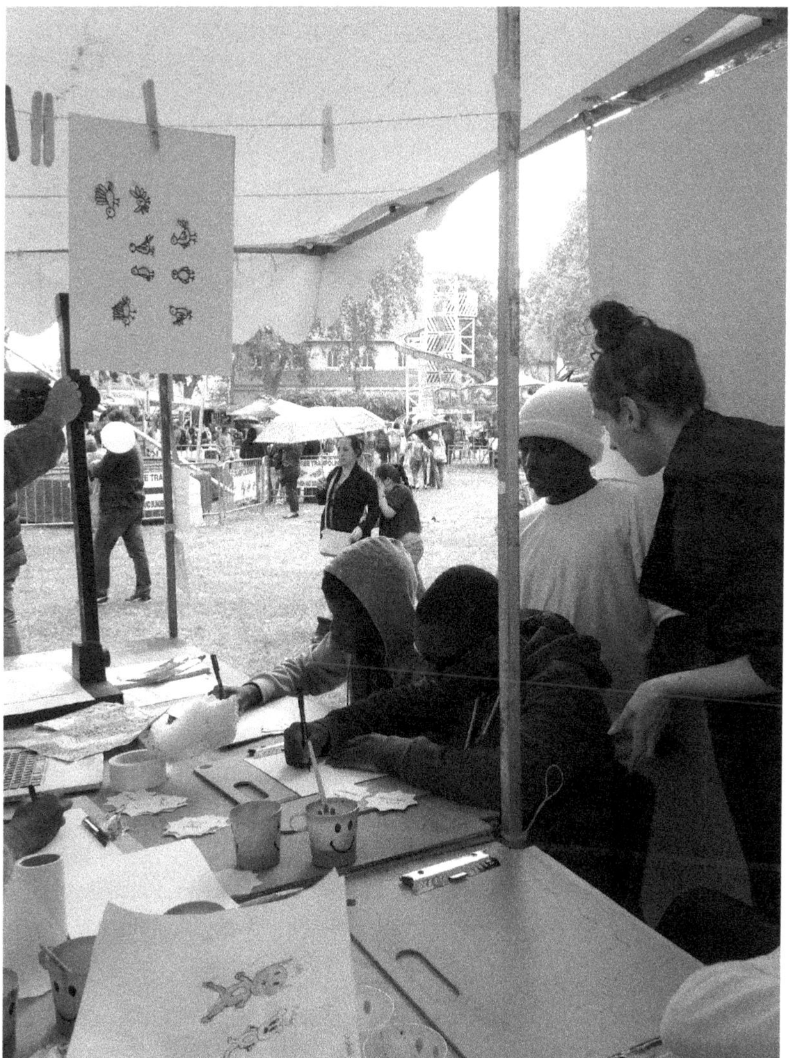

Figure 6.2 The *Play Your Place* stand at South West Fest 2014. Source: Harriet Hawkins and Ruth Catlow.

As local community members walk around the park where the festival stands have been set up, a number stop at the *Play Your Place* stand (Figure 6.2). The stand has been arranged with a series of seats at folding picnic tables to create drawing stations, each station with ready access to paper, paints, pencils. As they sit down participants are asked to think about an issue or a challenge they think their community faces. Included at the drawing stations are some prompts to get people thinking. Written on fluorescent card shapes notes such as 'place', 'actions', 'beings and things', 'the bigger picture' helps direct participants' attention. They offer a basic 'scaffolding', helping participants think about what to draw and about the elements of local experience that they want to explore further.

Often it is children who sit down first, some come over to see what is going on and attracted by the ongoing drawing and discussion sit down and begin to draw and chat. Adults linger, looking over shoulders at the drawings being done or listening to the discussions about the drawings and the linked community issues. One of the first to sit down is a young boy, together with his parents, and they begin to talk about what issues they have with where they live. They talk about a lack of open space, the boy wants a dog, and there is a discussion about how parks are disappearing (leaving less space to walk dogs). During their discussion, the boy decides to draw a dog in a park and as he starts to draw he becomes frustrated; his dog does not look like a dog.

Participant drawing is often used in situations where there is an assumption that participants will struggle to express ideas. This might involve young people or research contexts where there are cultural and linguistic differences, or it might concern things not easily put into words – expressive practices or lived experiences, for example. Across such discussions there seems to be an apparent assumption that drawing is somehow a more basic form of communication than words, or somehow more universal. It is, of course, important not to overlook issues raised by participants around not being good at drawing. As Young and Bennett note (2001: 145), literate children often perceive an artistic inadequacy, principally worried (like the boy with his dog) that their drawings do not look right. Furthermore, while drawing is often assumed to foster a certain form of attentiveness to that being drawn, frustrations at not being able to draw can break down that relationship (Hawkins 2015). Issues of inadequacy aside, not being good at drawing was not so important in the *Play Your Place* process. To focus on the quality of the output or the information that can be deduced from it is to overlook the importance of the drawing process.

As the afternoon progresses more and more people are attracted to the *Play Your Place* stand and the range of drawings grows. Dogs remain a theme, both having enough open space as well as having loving homes; Battersea Dogs Home is close by. For many residents, however, the issues they draw and explain open out a diverse range of lived experiences of the local area. The images show different levels of abstraction and representation, some are schematic or attempt a likeness, others are deliberately cartoonish, some have annotations. With various degrees of skill and experience participants draw and describe how government cuts are present in their daily lives and how gentrification is a felt force, as parks are closed and local services – nursery schools, local housing, toy libraries – are threatened whilst house prices rise, with the latter dramatically changing the make-up of local neighbourhoods, pushing out local families in favour of short-term renters often present only on weekdays.

Many of the drawings take an almost cartoonish form, for example one boy draws the glass-fronted expensive blocks of flats that have grown up in the local area as 'zombie buildings', describing them as descending from above, squashing what exists and taking over the area. Another man draws pianos and singers, reminiscing about the past as local pubs and bars are replaced by high-end outlets catering to lunching office workers and Monday-to-Friday renters who work in the nearby government buildings. These local issues play out against the backdrop of a series of geographically locatable drawn elements. Whether it be the recognizable mansions that form much of the area's privately owned residential stock, the iconic social housing blocks or key locations such as nearby Vauxhall Bridge, the Channel Four building or the Tate Gallery.

Drawings are often understood to possess a 'condensing' quality (Bagnoli 2009: 551), in this case narrating complex stories about the residents' plans, dreams, dilemmas and emotions. As such we witness the value of creative methods for enabling people to 'communicate in a meaningful way about their identities and experiences, through creatively making things themselves, and then reflecting upon what they have made' (Gauntlett and Holzwarth 2006: 83). There is more here, however, than good communication. As Gauntlett and Holzwarth (2006) note, drawing can be a useful activity for allowing participants time to reflect on issues being explored. As they note, many language-based methods rely on instant responses captured as data. They argue that drawing allows a reflective process, a taking of time, resulting in more thoughtful reflection. Asking people to draw offers a different way into a research question. For Gauntlett and Holzwarth (2006) it is the reflective process of making an artefact or a drawing combined with the embodied act of making something that you can look at, think about and change that marks out drawing and other creative processes from spoken responses in say interviews or focus groups. In the case of *Play Your Place* people made the drawings, often spending up to 30 minutes over them and then spent time with the group talking about the drawings and engaging with the wider debates. For Gauntlett and others what often seems to be of most value is what the researcher gains from the 'being attentive' associated with these drawing methods. What is perhaps more interesting to our discussion here is what this attentiveness means and does for the people involved in the process. It is not only that drawing enables the conveying of lived experiences, it can also enable our attunement to them, allowing, for example, for an exploration of the textures and characteristics of a place, a sense of engagement with a location or community that might pass un-thought or unremarked upon but which can often be deeply felt (Hawkins 2015).

Towards the end of the afternoon, as the festival starts to wind down, firemen and policemen who have been manning the event appear at the stall and start talking to the *Play Your Place* team. Initially unwilling to draw, one fireman is soon persuaded to sit down at the table and draw what he has been talking to the team about. He draws a closed-down fire station, as he explains that the Coalition Government has shut a whole series of fire stations across London, leaving many communities with reduced access to emergency services. In the normal course of the project, people who had done drawings would then describe their drawings for an overhead camera with an attached microphone that captured their voice-overs. Pointing to local buildings and important details in their pictures, residents narrated their sense of place and their issues, recounting the planes that flew overhead, the value of local parks, bars and shops, the persistence of some traditions and the demise of others. Prompted by questions from one of the team, they talked through what challenges they found in their local areas, as well as what futures they would like to see and the strategies by which they hoped these could be achieved. The next stage of the process involved those who wanted to be filmed against a white screen with a video camera holding their picture, creating a gallery of the local community and their images. The fireman, whilst happy to discuss his image and have his ideas recorded, was not willing to be identified or pictured with his drawing.

This process of talking through the drawings and of making a gallery of the participants with their drawings is a point at which the drawings become artefacts and information sources that are then used as part of the game-making process. As an information source, what is valued about these images are the lived experiences of place that these visual artefacts and their discussions offer. This accords with many of the participatory visual methods, during which a key point in the process is encouraging participants to explain their drawings. Drawings are seen as openers, as ice-breakers or 'tickets to talk' (Mitchell and Norgaad 2011). Interaction with produced images, whether diagrams, drawings or photos, provides a basis for engagement and communication between researcher and participants (Crilly, Blackwell and Clarkson 2006). What is often seen to be core to the value here is an attention to the meanings that participants themselves associate with the images they have created.

Whilst an important information source, once again we can appreciate what happens in the process of speaking and explaining as much as we can value the information gained. The drawing process and the subsequent explanation can be thought of as literally giving voice to and rendering visible the participants and their views – voice here having multiple meanings, in terms of both an expression of feeling or opinion and having the right or opportunity to express an opinion. As such *Play Your Place* offered the space to overcome silences, in both groups of people, and also in individuals, where elements of life might be sensitive and difficult to relate in words. This will be picked up in the following section of discussion on connecting communities, again thinking this through less in terms of the output and more in terms of the processes of encountering the art project as one of its participants.

What does drawing do?

To ask what drawing does is to query how embodied material practices shape both subjects and their ways of knowing. In *Play Your Place* drawing offers a data source, allowing insight into participants' lived experiences, bringing us closer to an understanding of the incredible complexity and diversity of the concrete realities of participants' lives, factors that come to inform the creation of the *Play Your Place* games. We might think then on the one hand about drawing as providing information for the games to be made, offering plots, characters and game elements. They are also a rendering visible and audible of the voices of often unheard political actors. On the other hand, what is equally interesting about these processes is that as well as constituting data, something also happens to those who participate in the production of these drawings (Bagnoli 2009, Morrow and Richards 1996). As participants, many of whom did not normally engage with the political process, thought about what to draw, sat talking to others on the table – both existing friends and new acquaintances – they were differently orientated towards their places. As they probed what they wanted to draw, reflected on their lived experiences of their communities – what had changed, what they already missed and what they feared losing – they became attentive to their local places in often new ways. Through the act of drawing, community members began to think of these places in terms of challenges, and produced a series of images that in the process of their making supported sense-making for individuals and collectives in the community. In summary then, the processes of drawing and the dialogue they promoted required participants to engage with the drawings they made, developing ideas and examining their stories, assumptions and community issues. It is also a process that re-orientates the participants towards their places, encouraging them to have an active role in shaping their futures.

Play Your Place 3: connecting communities – creating publics

Sitting around the table in the park in South Westminster, sometimes sheltering from the rain, people began to talk, sharing paint pots and coloured pens, discussing their drawings. Reflecting on what to picture, they shared stories as they waited for their turn to draw or to talk about their drawings. In common with digital storytelling, it could be argued that *Play Your Place*'s forms of collective game design and play offers a series of 'social goods'. The process of designing the games was incredibly interactive – from the collective brainstorming of ideas and small group discussions, to the drawing of the game elements, the processes of digitizing these game elements as well as sharing sound files, exploring existing game elements and trouble-shooting and strategizing issues. Not to mention, of course, playing the games when they were done. In these interactions, observed at a range of points in the process, the focus would gradually fall away from discussion of the local issues, to the emergence of an atmosphere marked by prosaic social interactions. The spaces of *Play Your Place*, whether outside in a park, in a village hall, or in a gallery would be filled with laughter, chat and exchange as attention shifted from the processes of drawing to more general social interactions, interactions that have proven benefits for happiness, confidence, self-esteem and the enhancement of social skills. Such ideas, however, do not exhaust the ways in which the social relations generated in the course of *Play Your Place* shaped the singular and collective political subjectivities and experiences of those involved.

Making, suggests Gauntlett (2011), is connecting. Art offers, McNally (2015) argues, a technology of connection. While Gauntlett's catchy phrase has come to be emblematic of the contemporary craft movement, the potential for creative practices to connect communities has long been acknowledged, but less often have the mechanics of this process been engaged and understood. As we can see from the discussion of the connections formed in the collective processes of drawing (above), making did seem to be a form of connecting. In order, however, to understand how in *Play Your Place* making might be connecting in a rather more nuanced way, we want to explore the social processes that happen in the context of the project and to understand its spatial enactments through ideas drawn from participatory mapping.

To do so we take our leave from two relevant dimensions of the participatory mapping literatures. First, how participatory mapping and the use of online media tools have been deemed to 'recognize tactics to discern resistant, alternative or suppressed articulations and spatial imaginaries that might otherwise be overlooked or deemed insignificant' (Elwood and Mitchell 2013: 286). Second, we are interested in how we might move beyond these more representational articulations to think through what happens in the processes of mapping; how, in short, *Play Your Place* functions as a site for assembling political subjects.

Mapping: connecting communities, assembling political subjects

Play Your Place's drawings and games very easily fit into the expanded register of maps and mapping practices that have become increasingly well recognized of late (Cosgrove 2005). These drawings and their assembly into games select out and simplify elements of the rich and complex lived experiences of the local community. Whether this be iconic buildings (the Tate Gallery, the social housing blocks) or local traditions (hanging washing between buildings, allotment gardens), these carto-visualizations assert what these places are like. As well as including visual symbols for local things actually under threat – fire stations, parks – these drawings and games are also 'visual synecdoches for broader socio-spatial changes wrought at the intersection of gentrification and austerity' (Elwood and Hawkins 2016). To think about these drawings and games as mappings is to take on board how maps are not only a visual representation of a place but are also bound up in the processes of producing the social, spatial and political. Following post-representational cartographies, we can think then of maps as both representational and generative; that is, as Dodge et al. (2009) explain, maps not only assert how the world is, but also how the world should be and, what is more, can help to bring worlds into being. To understand the *Play Your Place* games through this lens is to draw on lessons we can learn from literature on participatory mapping.

Participatory mapping has been understood as a means of constituting new spaces of civic engagement and resistance, in particular enabling 'the participation, influence and agency of less powerful actors as these groups create their own alternative spheres of declaration/engagement and cartographic praxis' (Elwood and Mitchell 2013: 276). The mappings of local places that *Play Your Place*'s drawn abstractions of the neighbourhood produced in response to key issues provided these alternative spaces and practices of engagement. Creating drawings and (later) games together, community members make visible very different experiences or understandings of the community, and articulate desired futures that may be contradictory or uneven in their effects on other people and places in the community (Elwood and Hawkins 2016).

In the development of drawings and games we find a transitioning of experiences of place from an individual, sometimes quite personal experience, to a shared understanding and a group narration of change. In terms borrowed from digital storytelling, this is the movement from a 'me story' – a personal issue – into a 'we story' and then finally into 'our story' (Lambert 2013). *Play Your Place* thus enabled a collectivization of a group of individuals around both specific issues but also around a general sense of there being issues within the community that need addressing.

Yet *Play Your Place* does not insist that differences within the community be resolved (or flattened) into a single shared representation of 'common' interests and concerns. Rather it enables communities to acknowledge and engage with issues of local tension, exploring these in playful and accessible ways. The format of the project seemed easily to make space for 'multiple perspectives and simultaneous existence of hyphenated identities' (Lambert 2013: ix), as well as for multiple different skill sets and levels of technological competence. The collective making of carto-visual representations thus involves, as Elwood and Hawkins (2016) note, working from different subject positions, experiences, knowledge and needs toward visualizations that materialize situated knowledge. This was less however about the assembling of a community in common and rather more about a dynamic process of community mapping that assembles subjects engaged in the negotiation of difference (Brown and Knopp 2008, Elwood and Mitchell 2013).

It was not just local communities that *Play Your Place* sought to connect, but the project had a wider aim to enable the views of overlooked groups to become visible in the planning process. As a range of creative projects currently being undertaken demonstrate, too often the planning processes and political choices regarding our places and communities are carried out based on conventional categories and distinctions made between 'expert' knowledge makers and 'lay' knowledge makers (Hawkins et al. 2015, Lane et al. 2011). While efforts are made to enable 'lay' voices to be heard, too often these efforts are limited and, furthermore, they situate 'lay' voices as less knowledgeable and engaged than 'expert' voices. We might therefore situate *Play Your Place* alongside other forms of public experiment, including using storytelling and workshops as part of local consultancy processes, wherein these public experiments become at the same time experiments in knowledge production (Kester 2004, Lambert 2013, Lane et al. 2011).

In its collectivized process of drawing, of making the games and their presentation on a website, *Play Your Place* has the potential to be a radical participatory method. We assert the radical here not just because, akin to discussions around participation in science, *Play Your Place* aligns lay knowledge alongside expert knowledge in a way that has the potential to shift the power relations in the political process. But because, following Lane et al. (2011), *Play Your Place* is as much about creating a new public capable of making a political intervention as it is about producing a solution.

This is to suggest, via the participatory mapping literatures, that *Play Your Place* might constitute its participants, often overlooked groups, as political subjects. As they sit at the table and debate the closure of fire stations, the decline in the amount and quality of public space, issues around schooling and housing, and the growth of second-home ownership in the area, the participants are created as politically engaged beings. Drawing their issues and then explaining them to the camera and to others can be understood as offering a 'pathway for formation of collectively held explanations and interpretative frames' (Elwood and Mitchell 2013: 283). As such, *Play Your Place* does not just constitute political engagement by actors from below, but evolves practices that constitute the critical antecedents to such action, creating subjects as critical thinkers, mobilized actors and active agents in collaborative knowledge-making processes.

What does mapping do?

To think about *Play Your Place* as about mapping is not only to appreciate the carto-visualizations of place that it proffers, but is also to use participatory mapping literatures to understand the politics of these individual and collective mappings. Making/Mapping in this case is connecting, coming to give voice to those groups often overlooked in the political process. More than this, however, *Play Your Place* might also be considered to 'foster the formation of political subjects, the formation of collective action frames (which might spur these political subjects to action), and collaborative formation of shared knowledge' (Elwood and Mitchell 2013: 277). As such, these practices are not just political engagements by actors from below, but are also a mechanism by which participants can come to understand themselves as political subjects, generating critical insights and collectively held interpretive frames that can mobilize collective action, or collaboratively develop shared knowledge (Elwood and Mitchell 2013: 286). Making is not just connecting here, rather it is to make and remake communities that are more politically engaged, that are one step closer to action.

Conclusion: if you draw, it will happen

Play Your Place is surrounded by the kinds of claims that have come to be associated with contemporary art: namely that participants, communities and places are transformed through their encounters with these art projects. As this discussion has shown, *Play Your Place* does in fact do many things. We have explored the effects of the artefacts produced, the games, the drawings and their possibilities as both information sources but also as mappings. We have asserted how, by making visible and audible the views and lived experiences of this divided community, *Play Your Place* responds to concerns that 'in the service of equality and social justice, we will continue to be pressed to ask how some voices continue to be unrepresented, muted, unheard, "voiced over" and silenced' (Luttrell and Chalfen 2010: 143, 200).

Alongside thinking about what it is the drawn and digital artefacts do, we have been perhaps more interested in the practices of drawing and mapping and what it is that happens in the course of these embodied, material and collective practices. The point here has been the processes, to understand how in thinking through the mapping and drawing practices we can comprehend how these art encounters shape subjects and connect communities. We might of course wish that we could draw a direct link between *Play Your Place* and changes to the place, whether this be effective challenges to the structural inequalities or a reversal of the decline of community spirit and the closure of services. We can't do that, not yet at least, in part because the project has not gone that far. What we can see, however, are more temporary and fleeting shifts and changes, in for example the temporary creation of community engagement and spirit through the drawing and mapping process, through the empowerment of individuals and the assembly of a group subject within the community that is invested in addressing local issues, at least for a certain period of time.

Often concerns with the effects of arts practices, like participatory methods more broadly, have focused around the need to 'scale' up. How, it is often asked, do such small-scale concerns, such small group intersections and effects change anything? What is it they do? We would answer here that to ask such questions is beside the point, it is to view ideas of transformation and change through a lens that overlooks the power in micro-politics. For at stake here are queries around what sort of politics and transformations participatory arts enable. Politics comes here to denote 'a range of individual and collective practices which act on or engage structurally mediated inequalities, the social and material relations of everyday life, and negotiations over identity' (Elwood and Mitchell 2013: 276). To think about micro-politics is to think about embodied engagement, material practices and affective encounters, it is to value and understand small-scale, individual and not necessarily permanent transformations as a form of politics. These politics are not somehow a lesser alternative, a trading down from other larger impacts and engagements not yet evidenced by *Play Your Place*, but are in and of themselves important ways of thinking about how encounters with art do work in the world.

Acknowledgements

The authors would like to thank Mary Flannagan, the PYP team, Rebecca Brookes from Tate Local and the South Westminster Community who participated in the development of *Play Your Place* and the South Westminster iteration described here. The authors also acknowledge the support of the AHRC/Creative Works funding and AHRC Cultural Value scheme that enabled this research to happen.

Notes

1 Cited on *Play Your Place* website [Online]. Available at: www.playyourplace.co.uk [accessed: 30 August 2015].
2 You too can play South Westminster online here: www.playyourplace.co.uk/southwest minster [accessed 30 August 2015].

References

Bagnoli, A. 2009. Beyond the standard interview: the use of graphic elicitation and arts-based methods. *Qualitative Research*, 9(5), 547–570.

Berger, J. 2007. *Berger On Drawing*. London: Occasional Press.

Bishop, C. 2012. *Artificial Hells*. London: Verso.

Bourriaud, N. 2002. *Relational Aesthetics*. Dijon: Les presses du réel.

Brown, M. and Knopp, L. 2008. Queering the map: the productive tensions of colliding epistemologies. *Annals of the Association of American Geographers*, 98, 40–58.

Cosgrove, D. 2005. Maps, mapping, modernity: art and cartography in the twentieth century. *Imago Mundi*, 57(1), 35–54.

Crilly, N., Blackwell, A. and Clarkson, P. 2006. Graphic elicitation: using research diagrams as interview stimuli. *Qualitative Research*, 6(3), 341–366.

Dodge, M., Kitchin, R. and Perkins C. (eds) 2009. *Rethinking Maps: New Frontiers of Cartographic Theory*. London: Routledge.

Elwood, S. and Hawkins, H. (2016) Visual politics: an intradisciplinary exploration. *Annals of the Association of American Geographers*, online first view.

Elwood, S. and Mitchell, K. 2013. Another politics is possible: neogeographies, visual spatial tactics and political formation. *Cartographica*, 48(4), 275–292.

Enright, E. and O'Sullivan, M. 2010. Producing different knowledge and producing knowledge differently: rethinking physical education research and practice through participatory visual methods. *Sport, Education and Society*, 17(1), 35–55.

Flanagan, M. 2009. *Critical Play: Radical Game Design*. Boston, MA: MIT Press.

Gauntlett, D. 2011. *Making Is Connecting*. London: Basic Books.

Gauntlett, D. and Holzwarth, P. 2006. Creative and visual methods for exploring identities. *Visual Studies*, 21(1), 82–91.

Hawkins, H. 2014. *For Creative Geographies, Geography, Visual Art and the Making of Worlds*. London: Routledge.

Hawkins, H. 2015. Creative geographic methods: composing, representing, intervening. *Cultural Geographies*, 22(2), 247–268.

Hawkins, H., Marston, S., Ingram, M. and Straughan, E. 2015. The arts of socio-ecological transformation. *Annals of the Association of American Geographers*, 105(2), 331–341.

Helguera, P. 2011. *Education for Socially Engaged Art*. London: Jorge Pinto Books.

Jackson, S. 2011. *Social Works: Performing Art, Supporting Publics*. New York and London: Routledge

Kesby, M. 2002. Participatory diagramming: deploying qualitative methods through an action research epistemology. *Area*, 32(4), 423–435.

Kester, G. 2004. *Conversation Pieces: Community and Communication in Modern Art.* Berkley, CA: University of California Press.

Kester, G. 2011. *The One and the Many. Contemporary Collaborative Art in a Global Context.* Durham, NC, and London: Duke University Press.

Lambert, J. 2013. *Digital Story-Telling: Capturing Lives, Creating Community.* London: Routledge.

Lane, S. et al. 2011. Doing flood risk science differently: an experiment in radical scientific method. *Transactions of the Institute of British Geographers*, 36(1), 15–36.

Luttrell, W. and Chalfen, R. 2010. Lifting up the voices of participatory research. *Visual Studies*, 25(3), 231–243.

McNally, D. 2015. Comforting others: sociality and the ethical aesthetics of being together, in *Geographical Aesthetics; Imagining Space, Staging Encounters*, edited by H. Hawkins and E. Straughan. Farnham: Ashgate, 178–200.

Mitchell, R. and Norgaad, M. 2011. Using DIY cartoon storyboards, live sketching and co-sketching to involve young and older users in participatory design. IASDR2011, I4th World Conference on Design research, Available at: www.researchgate.net/profile/Robb_Mitchell/publication/241685944_USING_DIY_CARTOON_STORYBOARDS_LIVE_SKETCHING_AND_CO-SKETCHING_TO_INVOLVE_YOUNG_AND_OLDER_USERS_IN_PARTICIPATORY_DESIGN/links/0deec51c9989ce2de1000000.pdf [accessed: 18 March 2016].

Morrow, V. and Richards, M. 1996. The ethics of social research with children: an overview. *Children and Society*, 10, 90–105.

Nieuwenhuys, O. 2004. Participatory action research in the majority world, in *Doing Research with Children and Young People*, edited by S. Fraser, V. Lewis, S. Ding, M. Kellett and C. Robinson. London, and Thousand Oaks, CA: Sage.

O'Neill, M. 2008. Transnational refuges: the transformative role of art. *Qualitative Social Research*, 9(2), 59–68.

Reason, P. 1998. Political, epistemological, ecological and spiritual dimensions of participation. *Studies in Cultures, Organizations and Societies*, 4, 247–267.

Tolia-Kelly, D. 2007. Participatory art, in *Participatory Action Research Approaches and Method: Connecting People, Participation and Place*, edited by S. Kindon, M. Kesby and R. Pain. London: Routledge, 132–140.

Toscano, A. 2009. The sensuous religion of the multitude: art and abstraction in Negri. *Third Text* 23(4), 369–382.

Young, L. and Barrett, H. 2001. Adapting visual methods: action research with Kampala street children. *Area*, 33(2), 141–152.

7 The production of temporary public space

Site-specific installation and 'vital materialities'

Gwen MacGregor

Introduction

For one night in late October, from dusk until dawn, the streets of downtown Toronto are turned over to contemporary art for the *Nuit Blanche*[1] festival. Over one million people bundle up and wander the streets to engage with hundreds of installations, videos, sculptures and performances that have been commissioned by the city of Toronto for the festival. It started in 2006 as a one-off event but has been embraced by city organizers and the public as an important annual contemporary art event. Much of the downtown is given over to *Nuit Blanche*; however, each year the organizers shift the areas in the city that are designated for the festival. This offers an opportunity for the artists and curators to engage differently with the specifics of neighbourhoods; it also shifts the burden that the impact of the more than one million viewers have on an area of the city. Parking lots, public swimming pools, bank lobbies, back alleys, major roads, public parks, cinemas and subway stations are just a few of the locations that have been used to present the work.

While the unprecedented audience numbers offer dynamic possibilities of interaction between the artworks and the viewer, the large physical scale and temporal brevity of the event presents some daunting challenges for the participating artists to create engaging and appropriate artworks. Increasingly there has been criticism from newspaper reporting and art press that much of the work resorts to hollow spectacle (Whyte 2010). Very little critical writing, however, has attempted a more detailed and nuanced critique of the highly varied experience of audience interaction with the artworks (MacGregor 2014).

In light of this, for this chapter I will critically engage with two artworks presented at Toronto's *Nuit Blanche* in 2013 and 2014. The first is a video projection, *La Conga Irreversible*, created by the Cuban collective Los Carpinteros[2] that I viewed at the 2014 iteration. The second is *The Little People*, an installation I co-created with the art and design collective Workparty,[3] that was presented in the Toronto City Hall in 2013. With respect to *The Little People* I will discuss the challenges of taking the work from the studio to the public arena of *Nuit Blanche* as co-creator. In contrast, the discussion of *La Conga Irreversible* will be strictly from my perspective as a viewer who saw it for the first time within the *Nuit*

Blanche context. While quite different in their physical form, both of these works used the public procession as a form of protest and celebration. I will call on 'new materialist' scholarship to address how the material agency and efficacy of these artworks can be understood within the unique and challenging environment of *Nuit Blanche*. It contributes to the body of work in this publication and elsewhere of scholarship (Connolly 2013, Hawkins et al. 2015) that discusses political practices and knowledge production across a range of dialectic sites, including the experience of both making and viewing public art as a complex multi-dimensional embodied process.

Before turning to a discussion of these works, I will outline the 'new materiality' literature relevant to understanding contemporary art production as both personal and political. It is important to acknowledge, however, that while 'new materiality' offers a productive shift in understanding the human/non-human relationship as not human-centred, it is limited by its tendency to default to the assumptive universal white subject. Thus, I will bring both black feminist and indigenous[4] scholarship into conversation with the 'new materiality' literature as both scholarships offer understandings that acknowledge the importance of recognizing the specificity of cosmologies, histories and lived experiences that in this case are relevant both to the making of the work itself, and to the multiplicity of meanings the artworks offer to members of the viewing public. The hope is that by bringing two artworks from the perspective of both artist and viewer into conversation with 'new materiality' scholarship, there can be a greater specific understanding of the complex proposition of a personal creative act that then shifts to become an object of public engagement. Can the artworks defy a static ontological categorization as spectacle and stand as vibrant objects that are intrinsically interconnected to both humans and human activity, while also showing the more-than-human creative process that artists engage with when making the work? At the same time, by recognizing the limiting capacities of 'vital materialities' to help answer this question – through a critique by alternative frameworks such as black feminist and indigenous scholarship – the complex, layered political potentiality of contemporary artworks exhibited in public space can be more fully acknowledged.

Interconnections with an artwork

Since 2010 several publications have emerged that develop possible understandings of 'new materiality' such as *Vibrant Matters* by Jane Bennett, *New Materialisms: Ontology, Agency and Politics* edited by Diana Coole and Samantha Frost (2010) and *The Fragility of Things* by William Connolly (2013). While none of these works engage directly with contemporary art to any large degree, they offer possibilities for a new perspective on biopolitics, political economy and more-than-human relationships that is a helpful and productive framework to engage with contemporary art objects. Like several other authors in the field, they rely heavily on Deleuze's critique of the organizational principle of dialectical materialism and instead consider a dynamic where 'difference is pushed to the limit, producing a shift to an affirmative relationality' (Deleuze 1968 [1994]). In the Deleuzian conception of matter, relationships in all their complexity and contradictions, and across different scales, precede human-centred activity. Bennett calls on Deleuze and Guattari's understanding of assemblage to put forward her ideas. 'Assemblages are living, throbbing confederations that are able to function despite the persistent presence of energies that confound them from within' (Bennett 2010: 23). Bennett describes this intricate responsiveness within complex alliances between all matter as a 'vital materialism'. She challenges the understanding in Western liberalism of humans as unique and as masters of nature and instead suggests that there is a material agency of non-human things that we are part of (Bennett 2011). This model intentionally shakes up our tendency towards human centredness as a way to reconsider our place in the world and our relationship to the things around us. Moreover, it proposes that it is the relationship within assemblages, and their capacities as much as their properties, that are most relevant to our understanding of the human/non-human connection (Dittmer 2013).

Peh Cheah (2010) has also drawn upon Deleuze's project to explore political thought, yet acknowledges the difficulty of imagining the political implications of this understanding of materiality as it is not derived from the labour of the sociohistorical subject so central to dialectical materialism. 'The figures [Deleuze] employs to characterize this power do not translate easily into our conventional vocabularies of political discourse and institutional practices' (Cheah 2010: 88). Cheah suggests that the answer may lie in questioning the understanding of the political itself. 'The effectivity of these materialisms lies in the urgency of rethinking the ontological bases of current languages and vocabularies of politics and political thought, beginning, for example, with the very idea of political organization' (Cheah 2010: 90). How does this relate to the two artworks in *Nuit Blanche*? The answer could be found by asking where the political is in the artworks. In 'new materiality' terms, it is not only in the subject matter of the work, but is embedded within the process of interaction between artist and the creative materials. Geographer Harriet Hawkins attends most affectively to this when stating:

> Much art made since [the twentieth century] has consciously enrolled explorations of artistic process in a series of critical dialogues surrounding the politics and places of arts production and consumption . . . It offers a means to access the sensuousness of urban living and a mode through which we can come to occupy urban spaces as sites of political responsibility and arenas for political engagement.
>
> (Hawkins 2010: 808)

The *process* of making an artwork is inherently implicated in its meaning and whether it is successful communicating to an audience.

Political theorist William Connolly takes up this line of enquiry in his book *The Fragility of Things* (2013), by suggesting that the relationship between nonliving nature and human beings can extend into the political world. Different degrees of 'creativity, nonhuman force fields and culture-nature imbrications' bring to the fore the gaps and dislocations neoliberalization makes on human activities and non-human fields (Connolly 2013: 10). Moreover, he suggests that they actually can help to extend the possibilities of our political sensibilities (2013: 13). His concern is that neoliberalism, with its focus on market forces, diverts attention away from other creative and natural processes and powers of organization that have nothing to do with monetary profit (2013: 11). An acknowledgement and development of human/non-human relationships pushes back against neoliberal capitalist mechanistic theories of causation (2013: 13). Connolly's notion of creativity is expansive, meant in the broadest sense, and these kinds of acts can and do take place throughout a range of human activity. It is not inappropriate, however, to use it more specifically in reference to both the creative act of making art and viewing art. In fact Jane Bennett has stated that the 'literary-visual-artistic ethical task' is to describe this human/non-human relationship (Bennett 2011). Artists know that 'fighting with' a material rather than 'working with it' can mean the difference between a successful or unsuccessful project. Following Cheah's re-examination of the 'political' and Connolly's proposition that the relationship of the living and non-living can extend into the political world, the creative act of the artist shared with a viewing public in itself can be understood as a political act. In a rather hard-to-define, Deleuzian moment, if the materials, the environment they are placed in, and the specificity of the moment, connect with a viewer the artwork and the viewer's experience come together in an ineffable moment that is greater than the sum of the parts.

While 'new materiality' and its Deleuzian roots are useful to consider this within a general framework, there is still an assumption of a universalist white subject that is experiencing the vitality. Even Connolly, who goes furthest to expand the understanding into a politicized sphere, still maintains a rather singular set of assumptions as to what the politics are and how best to respond. There are two bodies of scholarship that problematize the 'new materiality's' universality that are particularly productive in relation to the artworks from *Nuit Blanche* that I will be discussing. The first is by indigenous scholars such as Andrea Smith, Vanessa Watts and Arvin, Tuck and Morrill (Arvin, Tuck and Morrill 2013, Smith 2014, Watts 2013) who have been critical for the lack of acknowledgement of an understanding of human/non-human relationships that come from indigenous cosmologies. Watts, a Mohawk and Anishinaabe scholar, describes an Anishinaabe and Haudenosaunee cosmology that never created a hierarchy between human and non-human as is evident in Euro-Western epistemology-ontology (Watts 2013: 13).

> Place-Thought is a non-distinctive space where place and thought were never separated because they never could or can be separated. Place-Thought is based upon the premise that land is alive and thinking and that humans and non-humans derive agency through the extensions of these thoughts.
>
> (Watts 2013: 21)[5]

Fundamental to Watts' argument is that the Euro-Western scholars are not intentionally colonial, but that even with an acknowledgement of agency of the non-human within the epistemology-ontology framework, human and non-human are not addressed equally. The Euro-Western privileging of human as special dismisses an indigenous conception of the human/nature relationship, allowing colonial practices to occur (2013: 28).

There is another important body of work that helps not only to productively decentre 'new materiality's' universality but offers specific examples of new knowledge production that can emerge with the acknowledgement of multiple erased histories. Black feminist scholars such as Cecil Foster, Sarah Morgan Beckford, Katherine McKittrick and Charmaine Nelson challenge colonial narratives through an investigation into the history of the little-known Canadian slave trade (Beckford 2010, Foster 2010, McKittrick 2006, Nelson 2010). It was legal for over 200 years under various colonial regimes to own slaves in Canada. Most of the slaves were kept within domestic environments, leaving no lasting physical evidence equivalent to the plantations in the American south (Beckford 2010, Nelson 2016). In fact, the prevalent slave narrative that is most familiar to Canadians, and the one we are taught in school, is that Canada helped escaping slaves through the underground railway.[6] The black feminist scholarship not only problematizes a well-worn Canadian historical trope but, as will become evident in the next section, it demands an acknowledgment of multiple embedded histories within the physical environment that 'new materiality' does not attend to.

The artworks and their sites

I will discuss two artworks that have been shown in Toronto's *Nuit Blanche*, through the lens of 'new materiality' literature, as well as its problematization, to investigate its political potential. The installation *The Little People* was shown in 2013 and created by the Toronto-based collective Workparty, of which I am a member. The video projection *La Conga Irreversible* was created by the Cuban collective Los Carpentinos and was shown in 2014.

La Conga Irreversible

In 2014 the historic site of Fort York was used, for the first time, as one of the sites for *Nuit Blanche* and Cuban curator Magda Gonzales-Mora was commissioned to program the zone. It is an interesting choice of location with a layered and fraught history. Currently it is an historical park but it was established by British forces as a garrison in 1793 by Lieutenant-Governor John Graves Simcoe on the north shore of Lake Ontario, just west of the mouth of Garrison Creek (1787–1805 Plan of the Toronto Purchase). At the time, the land was home to the Mississaugas of the New Credit but it had also been used by indigenous peoples, including other groups of the Anishinaabe, Haudenosaunee and Wendat (Dragonfly 2012, Smith 2013, Stevens 2013, The Mississaugas 2012).[7] The 1600s and 1700s were a time of great unrest, war and death from disease for all the indigenous peoples as the colonizing Europeans pushed further west into their territories. The garrison was established primarily as a defence position against the United States but it was also the initial establishment of the permanent presence of Europeans in the region known first as the town called York and eventually in 1834 the City of Toronto (Geographical Name). The buildings that currently stand on the site, all built after the original buildings were destroyed in the Battle of York in 1813, have been used as a public museum since 1934. The museum only refers to the 'Mississaugas and Ojibways' in passing (Benn 2015) and does not mention the contestation of the use of the land by indigenous peoples, which continued until 2010 when the Government of Canada settled the claim by paying the Mississaugas $145 million for the land that is now Greater Toronto (Claim Settlement, Government of Canada 2010).[8]

Nuit Blanche curator Magda Gonzales-Mora entitled the exhibition 'Before Day Break' and included 12 site reflexive works that were placed on the 43-acre grounds. With a primarily international roster of artists, Gonzales-Mora was encouraging a reflection on Canada's history and, while not specifically mentioned in her curatorial statement, from the artworks presented it was clear that she was interested in engaging with the history of war and conflict within an international context (Scotiabank). One of the most striking and engaging works was a large video projection, *La Conga Irreversible*, created by the Cuban art collective Los Carpinteros. This work is a video documentation of a live performance for the *XI Bienal de La Habana* 2012 that consisted of 100 dancers performing the conga backwards down Paseo del Prado in Havana (Stolz 2013) – the music and the singing was also performed backwards. The conga dance originated in Santiago, Cuba as a street dance and has its roots in seventeenth-century cabildos [slave-nation societies]. At the end of the sugar harvest slaves were allowed to play music and dance in costume in the streets. There was an attempt between 1923 and 1932 by the mayor of Santiago to outlaw conga for its 'dangerous and contaminating morals' (Milstein 2014). For *La Conga Irreversible* (Figure 7.1), the dancers appeared out of a building at the top of the street but immediately started moving backwards down the partition in the middle of the boulevard. Instead of the traditional brightly coloured costumes they were dressed in black versions, made from recycled plastic bags and bottles. There were several hundred people in the street, mostly international visitors specifically there for the biennial. As the dancers started to perform the conga backwards down the street, local people began appearing out of buildings to watch. After about eight blocks the dancers turned around and came back but this time moved through the traffic, interrupting any possibility of normal traffic flow and eventually dispersed amongst the crowd.[9]

The Toronto iteration was not a live performance but a projection of the original event onto a 10 feet-high screen placed on the Fort York grounds specifically for *Nuit Blanche*. While the video had a clear physical separation from the audience, it is important that the work not be understood as a completely self-contained object in isolation from the surrounding environment. Placed next to one of the small garrison buildings the figures in the video appeared larger than life and

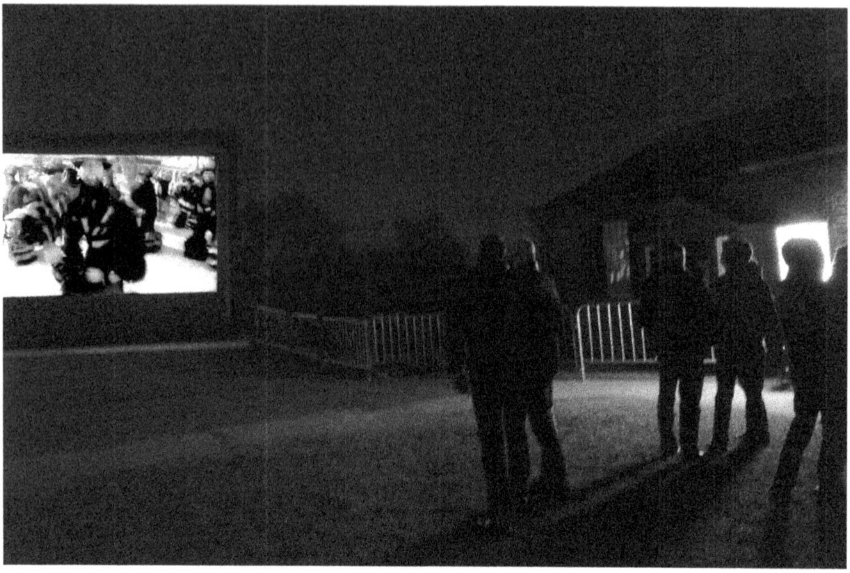

Figure 7.1 *La Conga Irreversible* by Los Carpinteros. Nuit Blanche 2014, Toronto. Source: Gwen MacGregor.

made the Fort York structures seem small and insignificant. The work became a complex combination of content and form that operated both within the frame as well as spilling onto the physical site.

To understand the full potential of *La Conga Irreversible*, its effect on the site, the viewers, as well as its political potential, it is useful here to return to the work of Deleuze. He wrote two volumes about cinema that explored the relationship between the moving image and thought (Deleuze 1968 [1994], Deleuze 1997b [1985]). He argued that cinema had the capacity to shock thought processes, forcing thought to think itself and disrupt chains of perception (Bogue 2003, Deleuze 1997a [1983]). He felt this was achieved by the fact that the image on the screen while literally contained on that screen, actually faces two directions – inwards towards itself and towards that which is outside the frame (Deleuze 1997b [1985]). The viewers' body and experience of the viewing is part of this, connecting with the image but also with their own memory and brain – a flow back and forth between image and thought across time and space (Rodowick 1997). With *La Conga Irreversible* the materiality was not embedded in a contained object; it moved between the image, the physical location of Fort York with its history, the physical buildings and the viewers themselves and their thought processes. These all worked together shifting across time and space to offer a dynamic assemblage of ideas, physical objects and people, that worked to prevent the artwork from becoming a static object. This dynamic offers a place where there is the power to shock thought (Deleuze 1968 [1994]) and where the political can survive and thrive. For it is in this creative in-between space that there can be contestation of the status quo and a possibility of nonviolent disruption (Connolly 2013). This suggests that video within an event such as *Nuit Blanche* doesn't necessarily have to be read as spectacle, if it is thoughtfully placed within an environment that can productively disrupt its spectacular tendencies.

It is here that the writing of the black feminists as well as the indigenous scholars can bring out the full political potential of *La Conga Irreversible* when shown in the Canadian context of Fort York. Indigenous scholars Arvin, Tuck and Morrill in the article 'Decolonizing Feminism' state that colonialism is a persistent political formation that exploits the land and the indigenous peoples and extracts value often through a system of slavery and labour exploitation (Arvin, Tuck and Morrill 2013). This links the practices of colonialism, slavery and exploitation of the land in the name of profit. Fort York begs the question, 'Were there any slaves at the fort at any point and was the garrison implicated in the larger slave trade that Canada participated in?' These questions are difficult to answer from historical records as so little of the history has been recorded (Beckford 2010, Cooper 2007, Foster 2010, McKittrick 2006, Nelson 2016). However, as slavery was not abolished until 1833, there could very well have been slaves on site. While the fort was built in relation to conflict with the United States, I became aware of a new tension while watching the black dancers in *La Conga Irreversible*, who seemed not only to challenge the Cuban government but the silence of Canadian history on selective issues. Now the Cuban dancers, projected on the large video screen next to the soldiers' barracks, seemed to mock the little brick buildings and refuse to go quietly into the oblivion of lost history. *La Conga Irreversible* reminds the viewer that the physical site of Fort York is not only a primary site of Settler Colonialism but a possible historic site of the slave trade. It also reminds us that while slavery is an historical event in Toronto it is interconnected to the present and persistent colonialism. 'New materiality' helps to remind us that the site is lively and interconnected to the artwork. The black feminist and indigenous scholarships demand that the liveliness must include and attend to multiple histories of colonialism and their ongoing presence.

The Little People

In 2012 I was invited by *Nuit Blanche* curator Ami Barak to participate the following year in his designated zone, Nathan Phillips Square, a large open space in front of the Toronto City Hall. I went during the 2012 iteration to look with fresh eyes, as an upcoming participant. When I arrived early in the evening the square was already packed with people; there was quite a festive air to the groups, consisting of a fairly young demographic who were chatting and taking photos. That year a large component of the exhibition was underground in the parking garage. I lined up with thousands of others, slowly inching down what normally would be the car ramp, to reach the underground parking lot. This experience gave me real trepidation about participating in the *Nuit Blanche* the following year, as it seemed that the sheer spectacle and scale of the event and a seemingly distracted audience overpowered any possibility of personal interaction with the artworks.

Barak is a seasoned Parisian curator who was the artistic director of the Paris version of *Nuit Blanche* in 2003 and 2004 (Ami Barak 2011). He is well acquainted with the possibilities and challenges that are particular to the festival and felt confident I could create an environment that would allow for relatively intimate moments for the viewers. His curatorial framing was to mark the 100th anniversary of Marcel Duchamp's first readymade, *Bicycle Wheel*, a bicycle wheel mounted on a stool (Rhodes 2013).[10] While I was excited by his proposal and his enthusiasm, I had difficulty imagining how to engage with the daunting context. Barak, however, was committed to presenting works that engaged with a material practice and he cited examples of my previous work that were heavily invested in physical materials, with hundreds of small ephemeral parts.[11] While conceptually the connection with Duchamp was understandable, I could not imagine how an installation of this type could survive the *Nuit Blanche* environment. Moreover, I had concerns because of the physical scale required to make a visual impact; even with a production budget, much of my work requires months of painstaking repetitive labour to complete.

Only after discussions with the collective Workparty, with whom I have collaborated with before, did it become clear that the answer was to invite them to work with me to create a project. Workparty is a loosely organized collective of designers and artists that collaboratively creates both design and art projects, including a diorama for a storefront that was an independent project for *Nuit Blanche* the year before.[12] The pooling of our intellectual and physical resources and the shared experience helped to embed the process of making into the artwork itself. The production of meaning could be built exponentially through our interactions and decision-making processes. In Connolly's terms of the political potentiality of creativity, we were embedding a politic of community and multiplicity in the making itself.

The group worked together for a year to create an installation called *The Little People* that was exhibited in the front foyer of City Hall (Figure 7.2). The concept was based on a real event that took place in Baraul, Russia, in 2012. Activists unhappy with corruption in a recent election were banned from protesting publicly. In response they substituted their own physical bodies with Kinder toys[13] and staged a toy demonstration. Unfortunately the toys were also banned by authorities who stated that inanimate playthings could not assemble for public political gatherings as they were not citizens (O'Flynn 2012). While the original protest was relatively small in size and number, Workparty scaled it up to a 30 feet-long installation that included 4,000 Kinder toys carrying protest signs.[14] We chose the City Hall foyer as it offered by far the best visual and political impact. The toys were placed starting just inside the front door and continued inwards in a straight line towards the council chamber. They were arranged in a long line on top of cardboard boxes obtained from the city, in a kind of evolutionary progression that created a narrative starting from the formation of creatures on the planet to the ultimate destruction of the environment and living things.

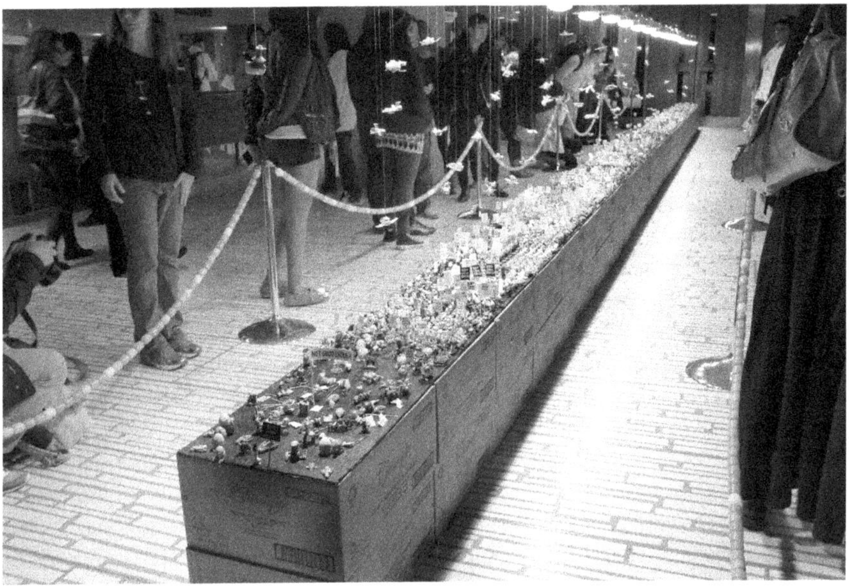

Figure 7.2 *The Little People* by Workparty. Nuit Blanche 2013, Toronto. Source: Gwen MacGregor.

Normally protests and placards are not allowed in Toronto's City Hall, and even our small Kinder toys had to obtain special permission. Their diminutive size and the shared understanding that they were harmless by Toronto city officials, unlike their Russian counterparts, allowed them to go where people could not. The group chose one single message that stated 'Not Good Enough'[15] to offer a clear statement of protest while intentionally not relating to a specific issue. In the context of the physical space of City Hall it could be read in reference to the turbulent situation being experienced by the local city council. At the time Toronto had a very controversial mayor who was known worldwide for his use of crack cocaine and for making bombastic, sexist, embarrassing statements to the press and public. This was not meant, however, to be the only reading of the work. The progression of the plastic toys as a procession through time, coming to an unfortunate environmental end, in combination with the 'Not Good Enough' message could be understood as an expression of the lack of government policy on climate change. The boxes the Kinders were sitting on had contained rolls of bathroom tissue purchased for use by City Hall employees, adding another layer of commentary on the state of affairs at City Hall. Each of the individual materials worked together to offer a strong clear message of protest while simultaneously retaining a sense of playfulness.

There are specific and interesting challenges related to creating a work for *Nuit Blanche*. Some are logistical, as many of the works are presented outdoors in spaces not usually equipped to exhibit artworks. The weather in October in Toronto is cold and can be very changeable. Positioning our installation in the indoor foyer eliminated these concerns and opened up the possibility for the use of more ephemeral materials. For an artist this can offer a greater freedom in the creation process, since materials such as cardboard and plasticine have an immediate responsive malleability that permanent materials often do not have. Management of crowds is also a real concern and one not familiar to most visual artists. A good day in a contemporary gallery can see a handful of visitors or hundreds, not hundreds of thousands. The first peak of visitors to view our work came at around 10pm and seemed to be the 'family crowd' and older adults. The second largest wave was at 1am and was a younger demographic, with groups of friends arriving from bars in the area. There was a stanchion as part of the work that functioned as a barrier for the audience but many people hung over it with arms extended, trying to take a photo or a selfie. There were moments when it seemed they would all tumble over the barrier landing face first into the work. Overall, however, there was a festive atmosphere with evidence of glee from much of the audience as they made personal connections with the materiality of the work. Many of the collective members stayed until dawn as we were mesmerized by the whole experience. When again would we have the chance to watch hundreds of thousands of people viewing and reacting to our work?

The relationship between the artwork as object, the artists who make it and the viewers is complex. There is an energy in the creation process that can be transferred to viewers through their interaction, articulated by Connolly as not entirely knowable. He describes the process as 'a world of becoming replete with multiple forces that sometimes intersect to throw something new into the world' (Connolly 2013: 133). Part of this process, however, is that it is difficult to know exactly how the work will 'behave' once it has left the studio. The artwork/audience interactions can be 'sometimes bumpy relations between real entities' (2013: 167). This is the power and risk of making art and putting it out into the world – it does not always behave.

The collective had hoped that the predominant response from the audience would be strongly weighted towards an identification with the art as a work of protest and dissatisfaction. The Kinders were standing in for the general public, protesting where people could not, hopefully creating a productive tension between the audience and the site of City Hall. However, the joyful festival atmosphere that comes with *Nuit Blanche* does not lend itself easily to a serious reflection of the work. At times the extent of the picture taking, for all members of Workparty, was surprisingly overwhelming — it was as if the work was being swallowed up by cameras, consumed and regurgitated as a digital image before our eyes. What we also had not accounted for was the extent of the strong nostalgic force that radiated from the Kinder toys. Certainly we were aware of it ourselves through the process of making the work; it was clear that many members of the public also knew and remembered these toys from childhood experiences. What we had not anticipated was that this would overwhelm other readings of the work. Time became an important element, even a material of the work, not the time that people spent looking at the work but the personal past that the viewers brought into the present and projected onto the objects (Connolly 2013: 88). One possible understanding is that the Kinder toys were not in fact 'vital' objects and instead remained fixed and inert, or at least 'resist' being repurposed by the artists. The fluid relationship between human and non-human materials expressed within the 'new materiality' literature may be somewhat inhibited with Kinder toys, because of their strong association to childhood consumer interactions. Considered in this way the relative vitality of objects needs to be considered when choosing them as possible creative materials.

Bruce Braun, drawing upon Bruno Latour, suggests that we have entered a period where our relationship to time has changed. Rather than facing the future we are focused on the past and unable to face the catastrophe before us (Braun 2015). Perhaps the nostalgic past that *The Little People* evoked was much easier to face than the possible catastrophic future it also suggested. Braun's statement, however, assumes a single past that we share that is in some way more comforting than the present. What if the past is painful or the information is unavailable? Are we talking about recent past or a much longer time period? Both the black feminist and indigenous scholarship is helpful to productively problematize the conflation of differentiated world views prevalent in Braun's statement. Black feminist scholars Beckford and Foster, for example, point to large gaps in the Canadian black history narrative. The black experience within Canada has been 'written marginal to the national discourse of belonging' (Beckford 2010: 282). This has to be acknowledged and the histories reclaimed for 'black people to lay claim to the Canadian nation-state' in the present (2015: 281) and shift Canadian norms and values that are still predominantly European (Foster 2015). In the case of *The Little People* did the Kinder toys with their singular message stand in too strongly for these norms? Indigenous scholarship suggests that this kind of questioning of accepted cultural norms is imperative to investigate the hegemony of knowledge production (Coulthard 2014, Hunt 2014, Tuhiwai Smith 1999, Smith 2014). Who is speaking in this installation and for whom? Furthermore, indigenous scholar Sarah Hunt suggests that to begin a process of decolonization for those of us who are not indigenous, we must recognize the limits of what we can know and accept there are things we cannot know. There is an important ethics to the recognition of 'not knowing' (Hunt 2014). These scholarships show that while the 'new materiality' is useful to question where the political is located in the artwork, it is through the engagement with the black feminist and indigenous literatures as critical methodologies that there is an opening up of the artwork to multiple, constructive critique (Ruddick 2010). The single message on the protest signs could easily allow the Kinders to stand in for a universalized white subject, prohibiting the greatest possible political expansiveness of the work. If we had multiple messages on the signs that the Kinder toys brought into City Hall that had been gleaned from individuals outside of the collective would the overall message of discontent be louder?[16] A greater awareness of these questions and the internal biases of the collective members could have brought a more effective attentiveness to multiple, layered possibilities for the artwork.

It is important to remember, however, that even with Workparty's experience of witnessing the audience viewing the work, that is not necessarily evidence of the entirety of their responses. Part of the work can unfold in the viewers' mind after the fact. It could have been that the strong nostalgic pull of the material presence of the Kinders was the first immediate response but that deeper, more lasting understandings of the installation could unfold over time. One of the great strengths of contemporary art is that it simultaneously uses different speeds and flow of time to deliver its messages (Deleuze 1968 [1994]).

Conclusion

I have been engaging with two installations that were presented in the unique and challenging environment of Toronto's *Nuit Blanche*. Both of these artworks used the public thoroughfare as a sight for marching and public demonstration, calling on a shared language and history of the physical act of political protest. Within that shared language each work offered specific strategies and physical materialities to address the challenge of engaging with a huge audience without resorting to the presentation of a mere spectacle. In its most vibrant form the exhibition of contemporary art in a public place encourages a co-mingling of audience and artwork, working to implicate the viewing public into the activity of the artwork.

'New materiality' frameworks have been engaged with to discuss *The Little People* and *La Conga Irreversible* to show the works' possibilities of political engagement through their materiality. It is an endeavour to express some of the subtle and complex relationships between artist and the making, the understanding of the power of the materials and the possibility for political engagement between artist, artwork and audience even at a large public event such as *Nuit Blanche*. There are limitations with this scholarship, however, as it does not allow for and fully acknowledge the particularities of language, communication and the different ways we may receive information given to us from an artwork. Nor does it really express the specificities of 'new materiality' – why are some materials more vital than others? Examples of black feminist and indigenous literatures help to parse out some of these shortcomings and give multiple specificities of meaning by effectively taking the 'new materiality' literature to task for not acknowledging black and indigenous histories and cosmologies. They also call for a greater awareness of ongoing colonial violence in academic methodological norms that continue to privilege a European universalized subject.

Ultimately viewing art remains a deeply personal experience and a viewer's connection to the work is a form of success irrespective of the intention of the artist. Sometimes for an artist the hardest part of the creation process is recognizing once it is public it is no longer theirs. Rather than falling into a trap of wishing for a specific takeaway it is more important to try and create works that generate a shared and affective environment that can offer multiple imaginaries (Hawkins et al. 2015). This comes in part from the efficacy of materials, their possible defiance of ontological categorization and how their multiple cultural resonances change in space and over time. It is here that there is political potential in understanding these public artworks, not with fixed meaning but open to in-between vibrant spaces of change.

Acknowledgements

I would like to thank Associate Professor Emily Gilbert from the University of Toronto for her help with this text. I would also like to acknowledge my graduate funding from the Social Sciences and Humanities Research Council in Canada.

Notes

1 *Nuit Blanche* originated in Paris in 2004 when museums and art galleries were open to visitors for one night. It has since become a worldwide phenomenon with Toronto offering the largest iteration (MacGregor 2014). From 2006 to 2015 the Toronto event was called *Scotiabank Nuit Blanche*. In February 2015 Scotiabank withdrew as the primary corporate sponsor. For this reason the festival will be referred to as *Nuit Blanche* for this chapter.

2 Members of the group are Marco Castillo and Dagobarto Rodriquez. There was a third member, Alexandre Arrechea, until 2003 when he left to pursue a solo career (Stolz 2013).

3 Workparty was founded in 2006 and has produced six projects. Participants for the *Nuit Blanche* 2013 installation were Mike Dudek, Bartosz Gawdzik, Gordon Hicks, Martina Hwang, Ryan leGassicke, Gwen MacGregor, Lewis Nicholson, Lena Rubisova and Emma Wright.

4 The term 'indigenous peoples' emerged in the 1970s from the American Indian Movement and the Canadian Indian Brotherhood. It internationalizes the issues and struggles of some of the world's colonized peoples (Tuhiwai Smith 1999).

5 Place-thought comes from Anishinaabe and Haudenosaunee creation histories that I cannot claim to have full understanding of and it would be inappropriate for me to retell as I come from a Euro-Western tradition. There are numerous indigenous scholars who have written about their cosmologies (Deloria 2003, Kimmerer 2013, Mohawk 2005, Simpson 2014).

6 At a talk at the Art Gallery of Ontario in Toronto, Associate Professor Charmain Nelson from McGill University in Montreal confirmed my own personal experience by stating that in the 10 years she has been teaching a Black History course, no student has come into the course knowing that there was black slavery in Canada (Nelson 2016).

7 The Anishinaabe were known as Ojibwa, the Wendat were known as the Huron and the Haudenosaunee as the Iroquois by the European colonizers. It is beyond the scope of this chapter to give a full recounting of the movement of indigenous peoples in southern Ontario pre-European contact, as well as early European contact. Details can be found at http://dragonflycanada.ca/toronto-is-an-iroquois-word.

8 In 1787 there was an agreement known as 'The Toronto Purchase' between the British Crown and the Mississaugas of the New Credit to purchase land. It was, however, contentious immediately after it was written for both representatives of the British Crown and the Mississaugas (Smith 2013). In 1986 the Mississaugas submitted a claim to the Indian Claims Commission based on evidence that there had been an illegal dispossession of Indian land and it was settled in 2010 (Treasury Board of Canada Secretariat 2007).

9 I did not see the original procession in Cuba. The description is from a firsthand recollection by Toronto artist Stan Denniston (2015).

10 The term readymade was first associated with artist Marcel Duchamp. Starting in 1914 he exhibited manufactured objects such as a bicycle wheel and urinal as stand-ins for painting (Duve 1991).
11 Examples of other installations can be found at www.gwenmacgregor.com.
12 For the 2012 project Workparty created a diorama that offered a dystopic view of the city. The artists that participated were Bonnie Devine, Mike Dudek, Gordon Hicks, Paul Kawai, Martine Hwang, Brianna Lowe, Gwen MacGregor, Lewis Nicholson, Lena Rubisova, Emma Wright and Ryan leGassicke.
13 A Kinder toy is a small plastic toy found inside a chocolate egg. They have to be assembled and include animals, cars, planes, figures. There are over 100,000 different designs.
14 Through social media we found a woman selling thousands of Kinder eggs without the chocolate.
15 Workparty was invited to show the installation again for a Calgary, Alberta event, Intersite Visual Arts Festival. Shortly before the installation was to be shipped we were informed that the organizers only wanted positive messages on the signs. This was unacceptable to the collective and we withdrew from the festival.
16 This question may have been attended to in part through a recent iteration *of The Little People* included in *Popup Downtown Regina*, September 2016. Workparty collaborated with Colonialism No More to share their messages of protest about government policy towards indigenous peoples. Messages such as 'Water is Life', 'Decolonize Gender' and 'Tipis Have No Closets' replaced the original message and seemed to resonate more loudly with the viewing public.

References

Ami Barak. 2011. Performing history. [Online: Performing History]. Available at: www.performinghistory.ro/ami-barak [accessed: 15 June 2015].

Arvin, M., Tuck, E. and Morrill, A., 2013. Decolonizing feminism: challenging connections between settler colonialism and heteropatriarchy. *Feminist Formations*, 25(1), 8–34.

Beckford S. 2010. Up rooting claims of legitimacy, in *Ebony Roots, Northern Soil: Perspectives on Blackness in Canada*, edited by C. Nelson. Newcastle upon Tyne: Cambridge Scholars Publishing, 277–296.

Benn, C. 2015. *A Brief History of Fort York*. [Online: *The Friends of Fort York*]. Toronto: The Friends of Fort York. Available at www.fortyork.ca/history-of-fort-york.html [accessed: 17 February 2016].

Bennett, J. 2010. *Vibrant Matter: A Political Ecology of Things*. Durham, NC: Duke University Press.

Bennett, J. 2011. Vibrant matter: zero landscape. [Online: *Eurozine*. 10 October]. Available at: www.eurozine.com/artiles/2011-10-19-loenhart-en.html [accessed: 10 August 2015].

Bogue, R. 2003. *Deleuze on Cinema*. London and New York: Routledge.

Braun, B. 2015. Futures: imagining socioecological transformation – an introduction. *Annals AAG*, 105(2), 239–243.

Cheah, P. 2010. Non-dialectical materialism, in *New Materialisms: Ontology, Agency and Politics*, edited by D. Coole and S. Frost. Durham, NC, and London: Duke University Press, 70–91.

Claim Settlement on the Horizon for Canada and the Mississaugas of the New Credit. 2010. [Online: *Government of Canada. Aboriginal Affairs and Northern Development Canada*]. Available at: http://news.gc.ca/web/article-en.do?nid=508179 [accessed: 10 August 2015].

Connolly, W. 2013. *The Fragility of Things: Self-Organizing Processes, Neoliberal Fantasies and Democratic Activism*. Durham, NC: Duke University Press.

Coole, D. and Frost, S. 2010. *New Materialisms: Ontology, Agency and Politics*, edited by D. Coole and S. Frost. Durham, NC, and London: Duke University Press.

Cooper, A. 2007. *The Hanging of Angelique: The Untold Story of Canadian Slavery and the Burning of Old Montreal*. Athens, GA: University of Georgia Press.

Coulthard, G. 2014. *Red Skin, White Masks Rejecting the Colonial Politics of Recognition*. Minnesota, MN: University of Minnesota Press.

Deleuze, G. 1994 [1968]. *Difference and Repetition*. New York: Columbia University Press.

Deleuze, G. 1997a [1983]. *Cinema 1: The Movement-Image*. Translated by Hugh Tomlinson and Barbara Habberjam. Minneapolis, MN: University of Minnesota Press.

Deleuze, G. 1997b [1985]. *Cinema 2: The Time-Image*. Translated by Hugh Tomlinson and Robert Galeta. Minneapolis, MN: University of Minnesota Press.

Deloria V. 2003. *God Is Red*. USA: Fulcrum Publishing.

Denniston, S. 2015. Artist. (Personal Communication, 11 March).

Dittmer, J. 2013. Geopolitical assemblages and complexity. *Progress in Human Geography*, 38(3), 385–401.

Dragonfly 2012. *Toronto Is an Iroquois Word*. Toronto: Dragonfly Canada. Available at: http://dragonflycanada.ca/toronto-is-an-iroquois-word [accessed: 1 October 2015].

Duve, T. 1991. *On Marcel Duchamp's Passage from Painting to the Readymade*. Minneapolis, MN: The Regents of the University of Minnesota.

Foster, C. 2010. The search for recognition and authenticity in Canada, in *Ebony Roots, Northern Soil: Perspectives on Blackness in Canada*, edited by C. Nelson. Newcastle upon Tyne: Cambridge Scholars Publishing.

Geographical name: the real story of how Toronto got its name. [Online: *Natural Resources Canada*]. Available at: www.nrcan.gc.ca/earth-sciences/geography/place-names/origins-geographical-names/9226 [accessed: 15 July 2015].

Hawkins, H. 2010. Turn your trash into … Rubbish, art and politics. Richard Wentworth's Geographical Imagination. *Social and Cultural Geography*, 11(8), 805–827.

Hawkins et al. 2015. The Art of Socioecological Tranformation. *Annals of the AAG*, 105(2), 331–241.

Historical Maps of Toronto. 2013. 1787–1805 Plan of the Toronto Purchase. [Online: *Historical Maps of Toronto*]. Available at: http://oldtorontomaps.blogspot.ca/2013/01/1787-1805-toronto-purchase.html [accessed: 7 August 2015].

Hunt, S. 2014. Ontologies of indigeneity: the politics of embodying a concept. *Cultural Geographies*, 21(1), 27–32.

Kimmerer, R. 2013. *Braiding Sweetgrass: Indigenous Wisdom, Scientific Knowledge and the Teachings of Plants*. Minnesota: Milkweed Editions.

MacGregor, G. 2014. The future of Toronto's *Nuit Blanche*. *Art & the Public Sphere*, 2(1–3), 103–116.

McKittrick, K. 2006. *Demonic Grounds: Black Women and the Cartographies of Struggle*. Minneapolis, MN and London: University of Minnesota Press.

Milstein, L. 2014. La Conga: Santiago de Cuba's Badge of Honor. *Revista*, 13(3), 40–42.

Mohawk, J. 2005. *Iroquois Creation Story*. Buffalo, NY: Mohawk Publishing.

Nelson, C. 2016. From African to Creole: examining Creolization through the art and fugitive slave advertisements of eighteenth- and nineteenth-century Canada and Jamaica. Presented at The Art Gallery of Ontario, Toronto.

Nelson, C. (ed.) 2010. *Ebony Roots, Northern Soil: Perspectives on Blackness in Canada*. Newcastle upon Tyne: Cambridge Scholars Publishing.

Nuit Blanche. [Online: *Nuit Blanche*]. Available at: www.scotiabanknuitblanche.ca [accessed: 12 May 2015].

O'Flynn, K. 2012. Toys cannot hold protest because they are not citizens of Russia, officials rule. [Online: *The Guardian*. 12 February]. Available at: www.theguardian.com/world/2012/feb/15/toys-protest-not-citizens-russia [accessed: 2 August 2015].

Rhodes, R. 2013. Ami Barak at Nuit Blanche: revisiting the readymade. [Online: *Canadian Art*, 27 September, 2013]. Available at: http://canadianart.ca/features/2013/09/27/ami-barak-nuit-blanche/ [accessed: 10 August 2015].

Rodowick, D. 1997. *Gilles Deleuze's Time Machine*. Durham, NC, and London: Duke University Press.

Ruddick, S. 2010. The politics of affect: Spinoza in the work of Negri and Deleuze. *Theory, Culture & Society*, 27(4), 21–45.

Simpson, A. 2014. *Mohawk Interruptus: Political Life across the Borders of Settler Sates*. Durham, NC: Duke University Press.

Smith, A. 2014. Native studies at the horizon of death: theorizing ethnographic entrapment and settler self-reflexivity, in *Theorizing Native Studies*, edited by A. Simpson and A. Smith. Durham, NC: Duke University Press, 207–234.

Smith, D. 2013. *Sacred Feathers*. Toronto: University of Toronto Press.

Smith, T. 1999. *Decolonizing Methodologies: Research and Indigenous Peoples*. London: Zed Books.

Stevens, S. 2013. The Historiography of New France and the legacy of Iroquois Internationalism. *Comparative American Studies* 11(2), 148–165.

Stolz, G. 2013. Dream logic, impossible combinations, and visual puns fill the work of Cuban sculpture duo Los Carpinteros. [Online: *Artnews*]. Available at: www.artnews.com/2013/06/24/los-carpinteros-profile/ [accessed on 10 May 2015].

The Mississaugas 2012. *A Sacred Trust*. [Online: *The Mississaugas of the New Credit Nation*]. Available at: www.youtube.com/watch?v=be1vfELwshg&feature=share [accessed: 4 November 2015].

Treasury Board of Canada Secretariat. 2007. *Indian Specific Claims Commission*. [Online: *Indian Specific Claims Commission*]. Available at: www.tbs-sct.gc.ca/rpp/2007-2008/iscc-crpi/iscc-crpi04-eng.asp [accessed: 23 June 2015].

Tuhiwai Smith, L. 1999. *Decolonizing Methodologies*. London: Zed Books.

Watts, Vanessa. 2013. Indigenous place-thought and agency amongst humans and non-humans (First Woman and Sky Woman go on a European world tour!). *Decolonization: Indigeneity, Education & Society*, 2(1), 20–34.

Whyte, M. 2010. Ennui Blanche? *The Toronto Star*, 4 October. Available at: www.torontostar.com [accessed: 10 August 2015].

8 'All your drains belong to us'

Young people and the non-representational geographies of public art in drain tunnels

Candice Boyd

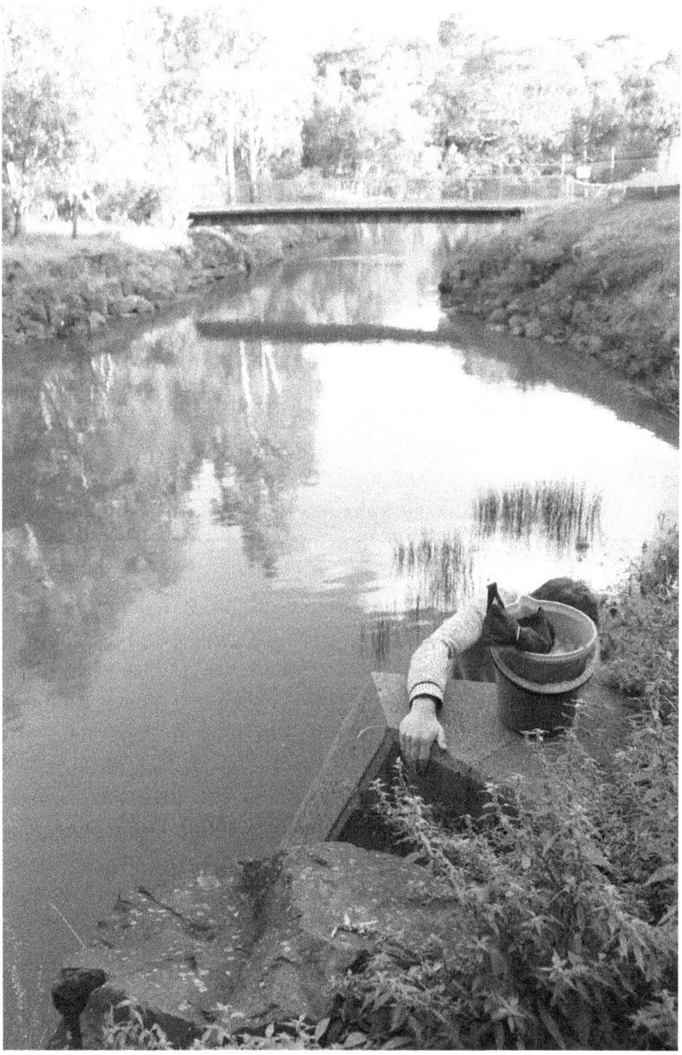

Figure 8.1 Descending into a drain tunnel. Source: Candice Boyd.

> The fleeting arthritis of cold and fear that gripped me slinks through, down the ridges of the chimney and into the ground, through the tunnels, into the water table. I'm limp, suspended, satiated.
>
> (Garrett 2014: 139)

> [T]he graffiti subculture [is] a 'liminal sphere'; a place where the features, ties and restraints of 'real life' are dissolved and new identity possibilities are offered up.
>
> (McDonald 2001: 9)

> The things we want are transformative, and we don't know or only think we know what is on the other side of that transformation. Love, wisdom, grace, inspiration – how do you go about finding these things that are in some ways about extending the boundaries of the self into unknown territory, about becoming someone else?
>
> (Solnit 2005: 5)

These excerpts of text present three contrasting dimensions of 'drain art' – the practice of graffiti in storm water tunnels. The first, articulated by urban explorer and human geographer Bradley L. Garrett, is the visceral experience of climbing above and under ground through otherwise abandoned 'public' places. The experience to which he refers operates internally and at the surface of the body as an intense, affective response to the overwhelming, yet thrilling, sensory demands of the spaces he encounters. It is effectively non-representational. The second dimension, as suggested by Nancy McDonald, is what some graffiti artists come to know by reflecting on their experience. It is thoughtful, cognitive, psychological, emotional and much easier to put into words than the first. The third by Rebecca Solnit is therapeutic and in sympathy with Whiteheadian and Deleuzian notions of becoming and difference. It refers to the transformative experience of getting lost – one that psychogeographers credit with the potential to 'suture up wounds' (Self and Steadman 2007). In this chapter, I will attempt to pay equal attention to each of these dimensions – the non-representational, the contemplative, and the therapeutic – in presenting an account of drain art practice and its limited public.

The majority of study into graffiti art and urban exploration has been ethnographic, using observation and in-depth interviews as 'data collection' techniques (e.g. Lee 2013, McDonald 2001, Rabine 2014, Taylor 2013). The account of drain art practice presented in this chapter employs a different, albeit kindred, method of inquiry. Informed by non-representational theories in geography and practice-led research in the creative arts, my research was performative rather than qualitative (Dewsbury 2010, Haseman 2010). As an artist-geographer, I practised what I researched and then translated that practice knowledge into creative outcomes. Thus, the chapter begins by briefly considering what it means to adopt a performative approach to researching public art encounters, before reviewing the literature on graffiti art practices. I then outline a period of fieldwork in drain tunnels of Melbourne (Figure 8.1), Australia, which incorporates a poetic account based on a soundscape produced from audio captured in the field. Reflections on the account in terms of its non-representational geographies are also offered. In bringing the chapter to a close, I share some insights into the 'therapeutic-ness' of drain environments, in reference to the range of peripheral practices that it encompasses and the needs of young people in an industrial age.

Performative research in human geography

Performative, non-representational and affect-based approaches to research in human geography take cues from practice-based methodologies in the creative arts by asserting process over outcome (Dewsbury 2010). Art is considered to be a 'way' that draws on human creativity to extend the boundaries of what counts as knowledge (Manning 2015). Thought is a mode of action that does not separate out the human subject from the ecologies of encounter. Rather, it attempts to retain its messiness by understanding subjects to be *internal* to experience and situated within a wider field of relations, 'a subjectivity that can never be disentangled from how the event came to fruition' (Manning 2015: 56).

Dewsbury (2003) mounts a case for 'knowledge without contemplation' in one of several papers on the implications of non-representational theory for geographical fieldwork. He describes it as 'the mix of emotions, desires, and intuitions within the aura of places, the communication of things and spaces, and the spirit of events'. There is a shared intensity in emotionally heightened spaces, which connects us to a wider ecology of things. As drain art practitioners, we 'lose' ourselves in the practice, but in doing so 'find ourselves in a buzzing world amid a democracy of fellow creatures' (Whitehead 1978: 50).

The aim of performative research in geography is to produce an account of the non-representational through a methodological practice that combines description and speculation (Anderson and Ash 2014). The emphasis is on 'background' and 'attentiveness' – a way of 'making sense' by cultivating an open attunement to 'spacings' in the fieldwork. Unlike ethnography, which invokes the reflexive self, performative research regards the self as geographically constituted and *distributed* amongst the myriad of relations between human and nonhuman of which the world is composed (Thrift 2008). Thus, research accounts produced in this way aim to foreground the background by adopting a style of writing which diminishes 'self as researcher' and elevates forces, flows and vital materialisms (see Bennett 2010, Bolt 2010; Boyd forthcoming).

Performative research in geography is also interested in affect – not only the way in which affect mediates spaces and places of encounter but how affect might be employed to 'enact the performance event again' (Dewsbury 2010). Thinking with affect re-orients inquiry to the dynamics of life; its intensities and capacities (Massumi 2002). An affective geography is a form of 'earth writing' that senses and expresses the 'aleatory dynamics of an impersonal, unqualifiable, life that exceeds processes of mediation or organisation and is only imperfectly suggested in the names that we habitually give to psychological emotions or bodily feelings' (Anderson 2014: 15). It senses the field of relations rather than the structure of relations.

Finally, performative research *presents* knowledges rather than interpreting, analysing or reducing them. This often necessitates a turn to creative praxis (Thrift 2008). Creative writing, film, soundscape, drawing or painting is employed not as mere documentation but as a creative effort to translate field-based experiences. Later in this chapter, I opt for three styles of presentation – the first is a de-centred poetic account, another is a soundscape made from naturalistic field recordings and the third is descriptive. Taken together, they are an attempt to 'reveal' the therapeutic-ness of drain art environments.

Graffiti as art practice

Graffiti is a site-specific practice to the extent that it is located and situational (Kwon 2002). Although more often identified with urban environments, graffiti practices also take place in rural and remote areas and on the peri-urban fringe (MacDowall 2014). Above ground, the height of a graffiti piece, not just its aesthetics, is related to its 'greatness' as it makes a statement about the lengths to which the artist was prepared to go (in terms of physical risk and risk of getting arrested) in the production of the piece (Taylor 2013). In contrast, the risk of getting caught practising graffiti in a drain tunnel is very low. Tunnels are rarely policed or surveilled. Compared to the 'showiness' of a lot of aboveground graffiti, graffiti in underground tunnels tends to be modest. The most elaborate pieces are around the 'mouth' of the drain where they can be seen by people on the 'outside'.

While the risks of being arrested are minimal, the physical risks for a drain art practitioner are considerable. The outlet of storm water drains are at the midway point of a river bed such that a practitioner gains entry by climbing down a rocky face, unsupported – risking broken bones, torn ligaments or head injuries in a fall as surfaces are wet and slippery. Inside a drain tunnel, it is completely dark (Figure 8.2). Practitioners wear inexpensive head torches that provide a limited light source while leaving hands free to carry buckets of paints and drip markers. Footing is unsure as the floor of the drain is curved and will often have water flowing through it. Furthermore, drains are 'stacked' so that progressing further along a drain involves climbing up poorly suspended ladders through subterranean waterfalls, materials in hand, while untreated storm water drenches clothing, making it difficult or impossible to see during the ascent. The most serious risk, however, is drowning should a storm water drain suddenly flood. For this reason, practitioners refrain from practising during periods of heavy rain. Unfortunately, this does not completely mitigate the risk (Dart 2008).

Modern-day graffiti practice is thought to have originated in Philadelphia in the 1950s, reaching new heights in New York City during the 1980s (Snyder 2011). While associated with hip-hop culture, Snyder (2011) notes that this was a later development and not a feature of early graffiti art. Many practitioners of graffiti identify as 'writers' rather than 'artists' (Abel and Buckley 1977). Older graffiti artists will often refer to a trajectory from 'tagging' as a teenager to 'throw

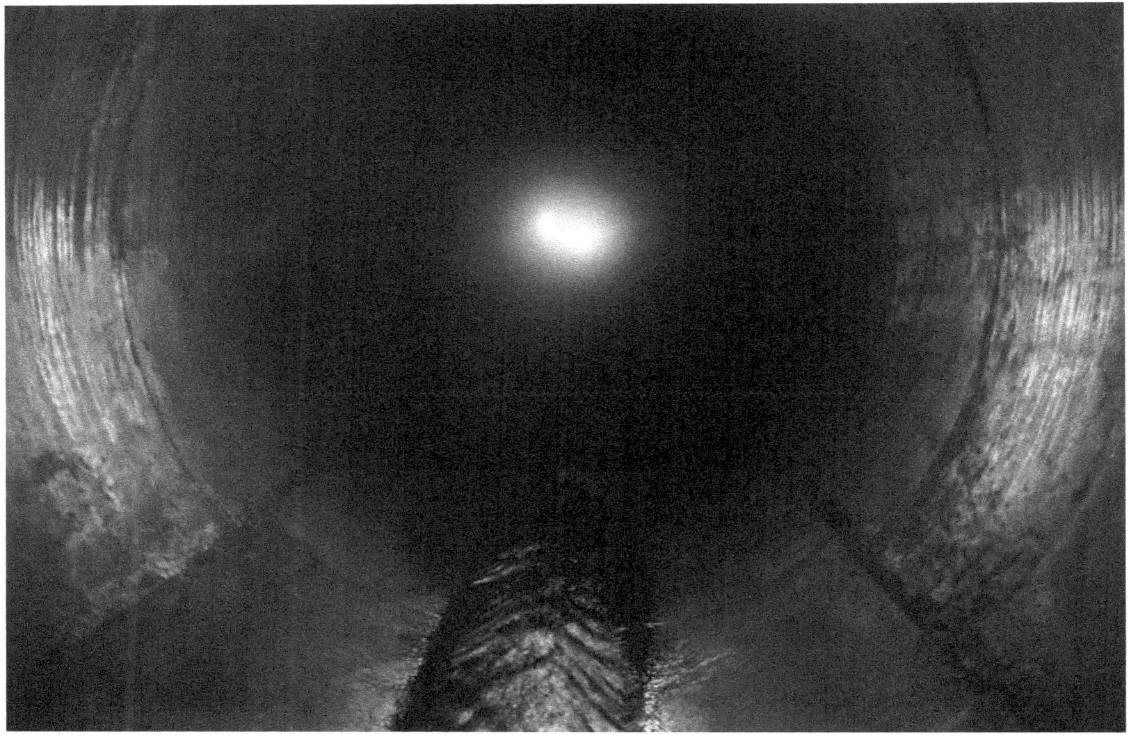

Figure 8.2 Inside a drain tunnel. Source: Candice Boyd.

ups' (employing a more intricate, bubble-style of lettering and the use of a single colour) to big and beautiful 'pieces' (McDonald 2001).

The elevation of graffiti to an art form came late in its history (Rabine 2014). Early writers in the arena of public art concentrated on the 'public art machine' – commissioned pieces by recognised artists chosen by elite committees or panels (Phillips 1988). Graffiti, although recognised as contributing to urban space, was regarded as vandalism. Bengsten (2013), however, notes that graffiti or street art shares many of the characteristics of public art such as its reliance on material context, its ability to spark controversy and public debate, and the production of everyday aesthetic encounters. Nonetheless, graffiti differs from mainstream public art in one important way – it is (for the most part) *uncommissioned* and *unsanctioned* (Young 2012). For this reason, Bengsten (2013) argues that it must be kept apart from the public art machine; if it is separated from its context, it loses its meaning.

Iveson (2013) cites graffiti and street art amongst a plethora of micro-political practices that are reshaping urban places – including guerilla gardening, flash mobbing, urban exploration and parkour. Although not always explicitly politically motivated, what these practices have in common is their enactment of 'rights to the city' and a presupposition of the equality of urban inhabitants. They have become a method of 'creating space within a space' by appropriating it for uses other than that for which architects and town planners originally intended. These actions have the potential to produce a more democratic society when it is those of lesser status who exercise this right. Similarly, Lee (2013) sees graffiti and street art as aesthetic empowerment, particularly amongst marginalised youth.

For young people, graffiti provides a unique mode of expression and a sense of belonging to a clearly defined subculture. As Rodriguez (2003) points out based on a review of ethnographic work, a key feature of graffiti is its anonymity. A young person can 'say' whatever they want to whomever they want on a wall without fear of retribution. As such, graffiti content may reflect anger at parental authority or society, political viewpoints, expressions of identity and group membership, or valued emotional connections such as friendship and romance. A less appreciated benefit, however, is an ethico-aesthetic one. The embodied practice of mark making provides a way of 'making sense' of the world. It presents an opportunity to experience the world materially and affectively – a characteristic of all therapeutic and transformative art practices (Collins 2014, Grosz 2009, McCormack 2013).

The fieldwork

The fieldwork for this project took place in 2013 as part of a wider exploration of the non-representational geographies of therapeutic art-making (Boyd 2016). Based on a series of case studies with a range of creative and spontaneous art practices, this work, in its entirety, attempted to think therapeutic activity at the boundary of the body and extending out, into the cosmos, rather than inward in support of a fragile ego. Informed by contemporary feminism (Braidotti 2006, Grosz 2009), Guattari's ethico-aesthetic paradigm (1995), Whitehead's process-oriented ontology (1978) and Deleuze's thinking on sense and the event (1990), the larger work sought to reclaim therapeutics as ecological, spatial and material.

Regarding the fieldwork that took place in storm water drains, I have a confession to make: I am no longer a 'young person'. At the time of conducting the fieldwork, I was 42 years old, married, with a 12-year-old daughter. I have, however, always had a degree of involvement with fringe communities, particularly mountainboarding (Boyd et al. 2009). In my experience, young people hold politically subversive individuals of any age in high regard. Therefore, there did not appear to be any impediment to engaging collaboratively with practitioners who were much younger than me.

Melbourne's network of drain tunnels is one of the most extensive in the world (Ryan 2014). Melbourne Water, the city's regulatory authority, maintains approximately 1,500 km (932 miles) of underground tunnels (McKernan 2013, Richter 2014). The 'Cave Clan' in Melbourne, a group of drain explorers established by three teenagers in the mid-1980s, initially shared information with other explorers by leaving messages on tunnel walls. The Cave Clan now have a website and a post office box, although they say they do not endorse accessing drains or give out any information as to the location of sites that are pictured on their website. Thus, it is still the case today that unless stumbled upon by accident, one needs to know someone who has already accessed a drain to know where it is. In my case, it was a 'friend of a friend' who introduced me to the first drain. Following two sessions of about four hours in one drain (known as the G.O.D. drain by Cave Clan), I was introduced to two other practitioners who took me to the ANZAC drain, home to the Cave Clan's ceremonial chamber, and 'Kent's Tomb'.

The fieldwork involved participating in several aspects of drain art practice, which apart from graffiti includes a range of peripheral practices including urban exploration, fire-setting, drain singing, height-related risk taking, public disturbance (subversion) and perspective-taking – all of which will be discussed in more detail later in the chapter. Nonetheless, for me these were research encounters. As a practice-led researcher, I was driven not only by the compulsion to create, but also to make use of the creative process to produce affective knowledges (Haseman 2010, Manning 2015). This involved a degree of openness to process and 'the event' as well as a need to gather material for future use in the studio. As such, I not only carried painting materials but also a camera, tripod and audio-recorder. The sensory material gathered in the field then formed the basis of a body of creative work. The wider project saw the creation of a short film, a painting, a photomontage, several digital art works and two sound pieces. By virtue of the drains' extraordinary acoustic properties, I took numerous audio-recordings during the fieldwork that were later used to produce a soundscape. The soundscape is housed on a companion website (https://soundcloud.com/dr-candice-boyd/all-your-drains-belong-to-us). The de-centred poetic account presented in the next section follows the soundscape's trajectory. It is recommended to listen first, listen to the soundscape, and then read.

Sounding drain art

> The bead in the can makes a 'thunking' noise as it hits the bottom and bounces back up again. Its hardness vibrates, creating the semblance of warmth against cold human hands. An index finger presses the nozzle and paint rushes out in a stuttering 'hiss', failing to coat wet concrete. Shake then spray. Repeat. It's not necessary to shake that much, but it sounds so good when the drain answers back.
>
> A human voice murmurs, its volume building as feet tread through water in search of company. A whistle pierces the droning, flirting with space. Human voices, human sounds, they can't compete with the roaring of storm water on its relentless passage from underground to overground. It drips, pours, ricochets, and tumbles; creating a fine mist that infiltrates the air.
>
> A spent can is thrown back into a bucket. It's difficult to see what choice to make – black, yellow, white, red. The head torch only illuminates so much in this dark, cavernous space. Improvise. The paint is dripping anyway. It's all wrong. Go for a wander with a drip marker instead. Looking out for past work, choose someone else's to write over. The drip marker scrapes against the rough concrete surface and splits, metallic purple spills out . . . ruined.
>
> Tracing the path of echoes leads to others. Laughter is heard as match after match is struck, each of them going out before they reach the water. Strike another one. A bit closer this time. Nervous. Sing from the throat and let it travel to the stomach. Create a harmonic, and feel the drain reverberate in return. That's really fucked up. Reminisce. Bust a rhyme. More wasted matches. Savour the anticipation of a massive burn. Are you ready?
>
> Our adventure ends, as the drain continues. Wade upstream, side-by-side, through stinking, icy water – resigned to wet, puckered feet. It's going to take a while, but who cares? Being nobody, going nowhere. Becoming alive. That's enough.

Experiencing the drain environment directly as a practitioner (or second-hand as a listener to a soundscape) enables us to appreciate how the ordering of the sense is challenged in this unique environment. Our sensorium is in excess of the optical in that we 'hear' so much more than we see (Hawkins 2014). Our experience is, at the same time, nonsensorial and immaterial – this is 'knowledge without contemplation'. It is the making of potential, which is energising (Deleuze and Guattari 1987, Whitehead 1967). We encounter the 'virtual' at the limit of process, felt in 'the arcing of experience toward the novelty of its taking final effect' (Massumi 2013: 17) – perceptually felt yet unseen and unheard, enveloping us and transforming us as we actualise it through material praxis.

The drain has ontology as an object (Harman 2012, Meillassoux 2013). As practitioners we only perceive the drain in relation to our own bodies but the drain also has a 'vastness', which extends beyond us. In 'the event' we experience this profoundly as a nonsenuous or amodal experience, leaving what Massumi (2013) describes as a nonconscious, operative 'trace'. It is one half of the double ordering of events that is a *nonlocal* linkage 'since all of the separate elements participate in it simultaneously from their individual positions . . . [d]irectly perceptually felt; "non-sensuously" perceived' (Massumi 2013: 107) – the spontaneous self-ordering of experience that gives form to life before there is a sense of self. Nonetheless, in these moments there is still a 'thisness' and a 'hereness and nowness' in which our individuality is expressed (Uhlmann 1996). Grosz (2005) describes it as a fracturing and opening up of the past which can bring forth the new: Deleuze (1990) as a point of inflection, or a singularity. It is also what we come to know through practice – a carnal knowledge that defies its description (Bolt and Barrett 2012). By 'sounding' drain art practice, instead of just writing about it, a sense of this space is re-created in an account of geographical fieldwork that attempts to do justice to its non-representational geographies.

Drain art and its peripheral practices

Although drain art is characterised by graffiti, it also incorporates a range of other activities performed in the spirit of urban exploration, psychogeography, and urban play (Garrett 2013, Landers 2012, Richardson 2015). One of these is height-related risk taking. As the opening quote to the chapter suggests, going underground requires 'making it' to an access point. Sometimes this involves climbing down cliff faces without support, sometimes it involves hanging off metal ladders that are loosely anchored by chains, and sometimes it involves swinging through gates or other openings. This aspect of the activity is a joy, not a burden. Practitioners may delay their entry into a drain for an hour or so as they revel in the sensations that accompany a body suspended in space.

Once inside the drain, there are other activities that provide breaks from painting and tagging. Perhaps the most dramatic of these is fire setting. This practice involves taking a container of petrol and slowly pouring it into flowing storm water. A volunteer in the group must then quickly drop a lit match into the stream of fuel and run in the opposite direction to avoid being burned. The result is a spectacular river of fire, which temporarily sucks the available oxygen out of the space and fills the drain with intense light. The fire is soon extinguished, returning the drain to complete darkness. Practitioners might play with different forms of display from firewalls to cascading 'waterfalls' of fire. For me, this was not only the most thrilling aspect of drain art practice but also one of the most social as the anticipation leading up to a fire being set, and the wonder as it burns, is a shared experience. At another level, the elemental nature of fire combined with a cavernous environment seems primitive, providing a point of contrast to life above ground (Glikson 2013).

A less risky practice associated with drain art is 'singing' or sound making. This involves the deliberate play of the human voice with the acoustics of the drain. Had I not approached drain art practice from the standpoint of a practice-led researcher, I might not have noticed the extent to which this form of play dominates the practice. On listening back to the numerous audio-recordings, it was apparent that the deliberate inflecting of voice happens from the moment people enter the drain until they leave. There are obvious examples in the recordings such as random shouts or whistles into the space, but there are also less conscious ones such as humming. On more than one occasion, a practitioner that I worked with spontaneously burst into song, inventing melodies without words that enhanced the resonant frequencies of the drain in some way. If another practitioner joined in, it might be to experiment with harmonics or pitch to create even more reverberation. As Reznikoff (2008) notes, large cavernous spaces (such as the Paleolithic painted caves in France) exhibit strong resonances which evoke intense emotional or spiritual reactions in people who encounter them. For me, the experience of 'drain singing' had an ethereal quality that provided a feeling of serenity and comfort.

Another peripheral activity, taking up the 'do-it-yourself' urbanism and right to the city that Iveson (2013) describes, is public disturbance. This takes place at the boundary of the inside and outside of the drain and may involve shouting or screaming through a grate to the 'world above' or physically jumping out of a 'manhole' to shock a random member of the public. On one occasion during my research, a group of us came upon a grate that opened out at a street-side dining area. Although we were unable to see the diners we could hear them talking and eating. One of us let out a piercing scream. The diners, not knowing where the sound was coming from, promptly stopped eating and talking. This made us laugh. On another occasion, we came out at a 'manhole' which could be opened and took turns in standing half in and half out of the hole, offering to shake hands with passersby.

The last peripheral activity I would like to highlight is perspective taking. There is opportunity for psychological reflection in almost every aspect of drain art practice as the environment constantly throws up challenges and contrasts compared to everyday life. As Ackermann (1996) argues, situated thinking allows young people to 'place' themselves in the world, negotiate meanings, and make their own ideas tangible. An example from my research was reaching the mouth of a very wide grate that opened out at the M1, Melbourne's busiest motorway. There was enough room at the mouth of the grate for four of us to lie on our backs looking up through the traffic to the clear, night's sky above. The bars of the grate were decorated with tags going back a full decade. We spent a long time there, in silence, staring at the night sky and listening to the roaring sounds of traffic as it passed over our bodies; mindful that the drivers were completely unaware that we were there. It was an embodied experience of reverie and a reminder of the need to affirm the 'value and charm of life' that has long been associated with the developmental stage of youth (Addams 2001: 1).

Conclusion

This chapter has embraced the promise of non-representational theory in geography to create alternative accounts of subterranean graffiti practice – a de-centred account, an example of 'sound writing', and a description of peripheral practices. Taken together, they reveal the complexities of the drain environment, its sonic power, and its potential for thrilling, daring and provocative encounters.

Bernard Stiegler (2010), who himself was jailed for juvenile delinquency, argues that the industrialised way of life inherited from the nineteenth and twentieth centuries has become 'toxic' for the spirit and the libido and that resisting it for the sake of future generations means cultivating new technologies of care. Based on the practice-led research in which I have engaged, I suggest that this is just what young people who practice public art in drain tunnels are doing. Unsanctioned and maintained by a limited public, the drain, with the multitude of ethico-aesthetic experiences it offers – perceptually felt or directly perceived – stands in stark contrast to the vicissitudes of young lives lived above ground.

A subject for further consideration is the level of 'psychic care' afforded by forms of creative expression which can only ever be viewed by a limited public. Storm water drains are prohibited environments not only by law but also by the degree of physical strength, balance, sure-footedness and stamina required to traverse them. This makes them places for the young and physically capable, for whom law breaking offers freedom. A sense of belonging exists which imparts a degree of safety in that expressions of angst, doubt, anger, prowess and ownership will be shared with a like-minded few.

The need to belong, the need to subvert, the need for adventure, the need for solace, the need for creative self-expression and the need for reverie are all met by the heightened affectivity and sensoriality of the drain environment. But perhaps more than any of these, the need for young people to claim some small part of the world as their own – and to 'write' their own geographies into it – is what makes this very specific form of public art so therapeutic for its practitioners.

Acknowledgements

My sincere thanks to the practitioners who introduced me to drain art practice although they cannot be named. Thanks also to Dr Bradley L. Garrett for his advice, mentorship, and encouragement since this research began; to Dr Rachel Hughes for expert supervision; and to the editors, Dr Martin Zebracki and Dr joni Palmer, for their on-going commitment to the scholarship of public art.

References

Abel, E. and Buckley, B. 1977. *The Handwriting on the Wall*. New Haven, CT: Greenwood Press.

Ackermann, E. 1996. Perspective-taking and object construction, in *Constructionism in Practice: Designing, Thinking, and Learning in a Digital World*. Mahwah, NJ: Lawrence Erlbaum, 23–37.

Addams, J. 2001. [reprinted from 1909]. *The Spirit of Youth and the City Streets*. Champaign, IL: University of Illinois Press.

Anderson, B. 2014. *Encountering Affect: Capacities, Apparatuses, Conditions*. Farnham: Ashgate.

Anderson, B. and Ash, J. 2014. Atmospheric Methods, in *Non-Representational Methodologies: Re-Envisioning Research*, edited by P. Vannini. London: Routledge.

Bengsten, P. 2013. Beyond the public art machine: a critical examination of street art as public art. *Konsthistorisk tidskrift/Journal of Art History*, 82(2), 63–80.

Bennett, J. 2010. *Vibrant Matter: A Political Ecology of Things*. Durham, NC: Duke University Press.

Bolt, B. 2010. The magic is in the handling, in *Practice as Research: Approaches to Creative Arts Enquiry*, edited by E. Barrett and B. Bolt. London: I.B. Tauris, 27–34.

Bolt, B. and Barrett, E. 2012. *Carnal Knowledge: Towards a 'New Materialism' Through the Arts*. London: I.B. Tauris.

Boyd, C. 2016. *Non-Representational Geographies of Therapeutic Art Making: Thinking through Practice*. London: Palgrave.

Boyd, C. Forthcoming. Forces, flows, and vital materialisms in therapeutic art making: a poetic account of contemporary geographical fieldwork. Under review with *ACME: An International Journal for Critical Geographies*.

Boyd, C., Kemp, E., Filiadis, T., Aisbett, D. and Markus, M. 2009. Mentors and mountainboarding: the development and delivery of an innovative program for rural adolescent males at risk of mental health problems. *Children Australia*, 34(2), 4–10.

Braidotti, R. 2006. *Transpositions: On Nomadic Ethics*. Cambridge: Polity.

Collins, L. 2014. *Making Sense: Art Practice and Transformative Therapeutics*. London: Bloomsbury.

Dart, J. 2008. Farewell for storm drain graffiti artist. *The Age Newspaper, 29 January*. [Online: The Age.Com]. Available at: www.theage.com.au/news/national/farewell-for-storm-drain-graffiti-artist/2008/01/29/1201369113190.html?s_cid=rss_news [accessed: 2 June 2015].

Deleuze, G. 1990. *The Logic of Sense*. [Translated by Mark Lester with Charles Stivale, Edited by Constantin V. Boundas]. New York: Columbia University Press.

Deleuze, G. and Guattari, F. 1987. *A Thousand Plateaus: Capitalism and Schizophrenia*. Minneapolis, MN: University of Minnesota Press.

Dewsbury, J. 2003. Witnessing space: 'knowledge without contemplation'. *Environment and Planning A*, 35, 1907–1932.

Dewsbury, J. 2010. Performative, non-representational, and affect-based research: seven junctions, in *The SAGE Handbook of Qualitative Geography*, edited by D. DeLyser, S. Herbert, S. Aitken, M. Crang, and L. McDowell. London: Sage, 321–334.

Garrett, B. 2013. *Explore Everything: Place-Hacking the City*. London: Verso.

Garrett, B. 2014. Battersea power station, in *Mount London: Ascents in the Vertical City*, edited by T. Chivers and M. Kratz. London: Penned in the Margins, 133–136.

Glikson, A. 2013. Fire and human evolution: the deep-time blueprints of the Anthropocene. *Anthropocene*, 3, 89–92.

Grosz, E. 2005. Bergson, Deleuze and the becoming of unbecoming. *parallax*, 11, 4–13.

Grosz, E. 2009. *Chaos, Territory Art: Deleuze and the Framing of the Earth*. New York: Colombia University Press.

Guattari, F. 1995. *Chaosmosis: An Ethico-Aesthetic Paradigm*. [Translated by Paul Bains and Justin Pefanis]. Bloomington, IN: Indiana University Press.

Harman, G. 2012. The Third Table/Der dritte Tisch: 100 Notes – 100 thoughts/ 100 Notizen – 100 Gedanken no. 85. *dOCUMENTA (13)*. Kassel: Erschienen im Hatje Cantz.

Haseman, B. 2010. Rupture and recognition: identifying the performative research paradigm, in *Practice as Research: Approaches to Creative Arts Enquiry*, edited by E. Barrett and B. Bolt. London: I.B. Tauris, 147–157.

Hawkins, H. 2014. *For Creative Geographies: Geography, Visual Arts, and the Making of Worlds*. New York: Routledge.
Iveson, K. 2013. Cities within the city: do-it-yourself urbanism and right to the city. *International Journal of Urban and Regional Research*, 37(3), 941–956.
Kwon, M. 2002. *One Place After Another: Site-Specific Art and Locational Identity*. Cambridge: The MIT Press.
Landers, F. 2012. Urban play: imaginatively responsible behavior as an alternative to neoliberalism. *The Arts in Psychotherapy*, 39, 201–205.
Lee, D. 2013, 'Anybody can do it': aesthetic empowerment, urban citizenships, and the naturalization of Indonesian graffiti and street art. *City and Society*, 25(3), 304–327.
MacDowall, L. 2014. GraffitiDownUnderGround. *Artlink*, 34(1), 18–21.
McCormack, D. 2013. *Refrains for Moving Bodies: Experience and Experiment in Affective Spaces*. Durham, NC: Duke University Press.
McDonald, N. 2001. *The Graffiti Subculture: Youth, Masculinity and Identity in London and New York*. Houndsmill: Palgrave Macmillan.
McKernan, M. 2013. *Underground Australia*. Canberra: National Library of Australia.
Manning, E. 2015. Against method, in *Non-Representational Methodologies: Re-Envisioning Research*, edited by P. Vannini. New York: Routledge, 52–71.
Massumi, B. 2002. *Parables for the Virtual: Movement, Affect, Sensation*. Durham, NC: Duke University Press.
Massumi, B. 2013. *Semblance and the Event: Activist Philosophy and the Occurrent Arts*. Cambridge, MA: MIT Press.
Meillassoux, Q. 2009. *After Finitude: An Essay on the Necessity of Contingency*. [Translated by Ray Brassier]. London: Continuum.
Phillips, P. 1988. Out of order: the public art machine. *Artforum*, 27, 92–96.
Rabine, L. 2014. 'These walls belong to everybody': the graffiti art movement in Dakar. *African Studies Quartley*, 14(3), 89–112.
Reznikoff, I. 2008. Sound resonance in prehistoric times: a study of Paleolithic painted caves and rocks. Paper presented at Acoustics '08 Paris, 29 June – 4 July, Université de Paris.
Richardson, T. 2015. *Walking Inside Out: Contemporary British Psychogeography*. London: Rowman and Littlefield.
Richter, D. 2014. Exploring the storm drains of Melbourne: a labyrinth of tunnels and creepy crawlies. [Online: Atlas Obscura.com]. Available at: www.atlasobscura.com/articles/exploring-the-storm-drains-of-melbourne-a-secret-labyrinth-of-tunnels-and-creepy-crawlies [accessed: 2 June 2015].
Rodriguez, A. 2003. Sense-making artifacts on the margins of cultural spaces, in *Expressions of Ethnography: Novel Approaches to Qualitative Methods*, edited by R. Clair. Albany, NY: State University of New York Press, 231–232.
Ryan, R. 2014. Melbourne's underground tunnel network is the most elaborate in the world. [Online: News.com]. Available at: www.news.com.au/travel/australian-holidays/melbournes-underground-tunnel-network-is-the-most-elaborate-in-the-world/story-e6frfq89-1227016196579 [accessed: 2 June 2015].
Self, W. and Steadman, R. 2007. *Psychogeography*. London: Bloomsbury.
Snyder, G. 2009. *Graffiti Lives: Beyond the Tag in New York's Urban Underground*. New York: NYU Press.
Solnit, R. 2005. *A Field Guide to Getting Lost*. New York: Penguin.
Stiegler, B. 2010. *Taking Care of Youth and the Generations*. [Translated by Stephen Barker]. Stanford, CA: Stanford University Press.

Taylor, M. 2013. Toward understanding the street code of silence that exists among prolific graffiti offenders. *Victims and Offenders*, 8, 185–208.

Thrift, N. 2008. I just don't know what got into me: where is the subject? *Subjectivity*, 22, 82–89.

Uhlmann, A. 1996. To have done with judgment: Beckett and Deleuze. *Substance*, 8, 110–113.

Whitehead, A. 1967. *Adventures of Ideas.* New York: Free Press.

Whitehead, A. 1978. *Process and Reality (corrected edition).* New York: Free Press.

Young, A. 2012. Criminal images: the affective judgment of street and graffiti art. *Crime Media Culture*, 8, 297–314.

Part III

Diversity

9 Mobilising the 'Right to Remain' in Vancouver's Paueru-gai

An art-based participatory research intervention

Aaron Franks, Jeff Masuda, Audrey Kobayashi and the Right to Remain Community Fair Team[1]

The Right to Remain Community Fair ('the Fair') was an art-based participatory research intervention in support of urban activists against the hyper-gentrification currently sweeping a low-income neighbourhood in Vancouver, Canada. The area is part of the Downtown Eastside (DTES), referred to as Paueru-gai[2] by the Japanese-Canadian community that once lived there – and going by less salubrious names among many city residents. The Fair's objective was to provide a stable but flexible platform for the current low-income community to put into 'aesthetic action' (Robinson and Martin 2016) their collective analysis of their Right to Remain (Franks et al. 2015a; NNM 2015) in this contested area.

The Fair, which ran July 2014–January 2016, was part of the multi-year project 'Revitalizing Japantown?', which addresses gentrification and dispossession in the DTES and calls into question recent planning initiatives that threaten to reimagine the neighbourhood into an upscale landscape boasting the trope of an exotified Japanese-Canadian community. Our focus here is on today's residents of the DTES, for whom the Fair intervened as resistant public art. We begin with a historical background on the neighbourhood and the research partnership, whose team comprised academics, artists, DTES inhabitants and allies. We then address the conceptual dimensions of the project that helped the research team bridge theory and practice: that is, a materialist reading of human rights that cuts across and links what are often presented as the chronologically and culturally separate struggles of racialized and marginalized inhabitants of the DTES; and a materialist reading of culture that indicts prevailing planning processes that separate activities identified as cultural from economic and social needs-meeting practices, paving the way for cultural commodification and the enclosure of public urban space. Finally, we address how the Fair engaged activist participants in a range of creative venues.

The Fair pushed back at the hyper-gentrification of the DTES to counter the whitewashing and fragmentation of the legacy of continuous community for racialized and other marginalized inhabitants. We worked through three general modes: provoking with artistic *form*; contesting and *spatializing* justice; and enabling *dialogue* within sites of encounter among diverse participants, residents and neighbourhood visitors, and facilitated through the Fair in numerous community spaces via workshops, walking tours, gallery exhibits and museums. Given the long-term and intimate involvement of the Fair's peer-artist facilitators, our analysis draws heavily from their experience and views.

Placing the story

Before the arrival of Europeans in the 1860s, *K'emk'emeláý*, or 'Place where maples grow,' in the language of the Indigenous Sk̲wx̲wú7mesh (Squamish) First Nation, had been an important meeting place for over 8,000 years. The Europeans logged its rich forests and dispossessed Indigenous peoples of lands that were never formally gifted or sold to the British Crown. It became 'Vancouver'[3] in 1886. A year later, the national railway arrived, linking the city to the lumber-hungry markets of industrializing North America. Like most of British Columbia, Vancouver still exists on unceded territory, in this case Sk̲wx̲wú7mesh (Squamish), xʷməθkwəy̓əm (Musqueam) and səl̓ílwəta (Tsleil-Waututh).

The neighbourhood

During the 1880s, a small community of almost entirely male Chinese immigrants who had worked on the railway settled in the southern part of the DTES. Their growth was halted by the Chinese Immigration Act (the infamous 'head tax') of 1885, and thereafter the Chinese-Canadian community struggled in low-paid service jobs: cooking, cleaning and laundering for the city's white inhabitants. By the 1890s, Japanese immigrants began to work sawmills along the south shoreline. The adjacent enclave around Powell and Alexander Streets, originally built by middle-class white families who by the turn of the century were migrating to newer residential areas, became inhabited almost exclusively by Japanese immigrants. They built dozens of boarding-houses for the millworkers, many of which remain today: creating an island of single room occupancy (SRO) residences surrounded by Canada's hottest housing market. Throughout the twentieth century, the area between these two Asian immigrant communities became increasingly impoverished and economically marginalized, known as 'Skid Road'[4] to the transient workers who congregated there.

Already tense labour divisions became broader resentment as European immigrants arrived in increasing numbers, heightening the racist rhetoric of British Columbia as 'White Man's Country' (Adachi 1976, Miki 2004, Ward 1978) and sparking sporadic violence and an anti-Asian riot in 1907. After a Royal Commission led to escalation of the 1885 Act against Chinese immigrants, and the Hayashi-Lemieux ('Gentlemen's') Agreement (1908) curtailing Japanese immigration, these nascent communities continued as ghettoized neighbourhoods in a racist society.

Paueru-gai grew over the early twentieth century to become a viable community. Several generations worked and lived in the neighbourhood, the de facto hub of a Japanese-Canadian culture that stretched up and down the West Coast. In 1942 over 20,000 Japanese Canadians were forcibly uprooted, dispossessed of property and citizenship rights and sent to internment or forced labour camps. The cost to the uprooted was immeasurable – health, infrastructure, livelihoods, education and cultural resources were all deeply damaged. Their properties and possessions, supposed to be held in trust by the state during the uprooting (1942–1949), were instead confiscated, held by the city or sold privately at bargain prices (Adachi 1976).

After the uprooting the DTES became a place of even greater segregation and public disapproval. Whether wilfully ignored by Vancouverites, preyed upon by opportunistic profiteers or marked as the place of unfortunate but ultimately undeserving poor, the DTES is represented as 'our nation's slum' (*Globe and Mail* 2009), a 'gulag' (Hopper 2014), and Canada's 'poorest postal code' (Christoff and Kalache 2007). Unchecked rent-seeking and speculation, gutted health services, racialized poverty, abrogation of social housing, discriminatory policing and ongoing cultural genocide of First Peoples are deemed 'regrettable' aspects of necessary urban progress. But the links between successive human rights abuses and the unbroken connection of all communities to the forces of settler colonization, systemic racism and class prejudice are largely ignored. This is reflected both in the deteriorating physical landscape of SROs occupied by successive groups of urban poor, and in the systemic failure to address either the material deprivation or the historical abuses and violations of human rights. Our research examines how the current inhabitants of the DTES are being targeted by gentrification under a new moniker – 'Japantown' – in an attempt to cleanse the neighbourhood's deeply troubled past through racialized urban branding.

Development of a Community Research Partnership

Revitalizing 'Japantown'? (RJ, 2012–2015) is a research partnership that supports community organizations, artists and researchers in challenging the fragmented, moralistic ways in which the neighbourhood is generally upheld. It extends our art-based participatory work in the community since 2007 (Masuda and Crabtree 2010) to promote understanding of urban inequities (Masuda et al. 2014, Teelucksingh and Masuda 2014). We challenge the moniker 'Japantown' as racialized, exotified, Orientalized and inaccurate, yet increasingly deployed *from the outside* as a trope to justify a commodified reinterpretation of Japanese-Canadian efforts to re-establish a sense of place in the neighbourhood (Takeuchi 1982).

This imposed, inappropriate branding has damaged the community's access to culturally appropriate resources, including health and social services, and its ability to exercise rights within the political sphere. The standards of development to which the community has aspired – a strong social economy, cohesion through diversity, mutual aid networks, vibrant common spaces – have been ignored or denigrated by planning, health and other authorities who adopt a 'deficit model' approach.

Dominant historical representations have also used deficit-based rhetoric, focusing on abandonment, absence, deprivation and depravity with a presentism that evades accountability or redress for ongoing conditions rooted in past cycles of dispossession. Such a framing is used to justify myriad forms of neighbourhood cleansing and renewal. Indeed, early overtures between the City of Vancouver and Japanese-Canadian advocacy groups[5] regarding heritage restoration made clear that reasserting a Japanese-Canadian presence would first require the cleansing of 'problems which continue to compromise the area. These are the high concentration of persons with social, alcohol and drug problems, vandalism, large volume of through traffic, lack of adequate parking and construction of large-scale institutions' (Takeuchi 1982).

Dominant readings of health and history have also denied the DTES as an authentic place through many gaps, elisions and inaccuracies. The basic details of the story itself have been abstracted and circumscribed, the narrative equivalent of Lefebvre's (1992) abstract space of urban governance, distributing an order of accumulation throughout the complexity of needs-meeting, dwelling and social encounter in city life. Such abstract spaces are composed and imposed in many ways, including practices of naming, zoning and regulation. The DTES has recently seen an amplification and spatial concentration of such practices: for example, the imposing of colonizer names like 'Pioneer Park' on public spaces in unceded Indigenous territory, still heavily used by Indigenous peoples. The name of 'Japantown' is a prime example of a manufactured concept: barring one newspaper reference in 1941 (Malone), the area was known to its Japanese-Canadian inhabitants as 'Powell Street' – or Paueru-gai. In adjacent Chinatown, which for over a century featured distinctive low-rise buildings, the removal of restrictions is destroying the identity of that historical landscape through high-priced condominium developments (Mickleburgh 2013). Properties containing businesses that have long served low-income residents are also being re-zoned for restaurants and other middle-class amenities ('John', DTES resident).[6]

The experiences of our interview participants and artist-researchers resident in the DTES suggest that marketization and social control are accelerating, and are increasingly orchestrated to facilitate only nominally participatory planning processes, such as the Local Area Planning Process for the Downtown Eastside (LAPP 2012–2014 (City of Vancouver 2014)), providing cover for a declining interest in urban health inequalities and a growing movement to establish a cultural spectacle of 'Japantown'.

The Revitalizing 'Japantown'? community–academic research partnership was formed to counter these trends, comprising eight DTES and Japanese-Canadian advocacy organizations working with academic investigators and students (Table 9.1). The Right to Remain Community Fair emerged as an element of this partnership.

Table 9.1 RJ-RRCF partners[7]

Project Partner	*Primary Role*[8]	*Location*
Gallery Gachet	Lead organization for RRCF; exhibition host; project advisor; community liaison; event venue; meeting space	DTES
Greater Vancouver Japanese Canadian Cultural Association	Liaison with Japanese Canadian community	Burnaby
Nikkei National Museum and Cultural Centre	Exhibition host; archival resources; liaison with Japanese Canadian community event venue; meeting space	Burnaby
PACE	Assistance with interviews	DTES
Potluck Café Society	Assistance with interviews; interview venue; food; community liaison; advisor re: Chinese community	DTES
Powell Street Festival	Event host; meeting space; project student hosting; liaison with Japanese Canadian community	DTES/ Powell Grounds
Strathcona Business Improvement Association[9]	Dialogue with business development interests	Strathcona/ DTES
Vancouver Japanese Language School and Japanese Hall	Interview and event venue; community advisor; liaison with Japanese Canadian community; project student hosting	DTES

The DTES/Paueru-gai, 'Superdiversity' and the Right to Remain

Superdiversity emerged a decade ago as an explanatory concept for demographic change in the UK. In Vertovec's (2007: 1) words, it is 'a notion intended to underline a level and kind of complexity surpassing anything previously experienced'.[10] The concept is closely connected to migration studies (Humphris 2015: 1) and metropolitan 'multiculture' (Berg and Sigona 2013).

In the DTES, superdiversity is a product of the struggle for in/visibility by peoples of vastly different 'tenures' of inhabitation in the same urban space. In a contemporary *episteme* highly attentive to mobilities (Cresswell 2006) and globalized urban capital (Massey 2007), the Right to Remain as articulated by our research participants reflects the continuous relationships of colonization, class stratification and 'othering' shared among very long-term neighbours. These include the more-than-symbolic 8,000-year-old presence of Sḵwx̱wú7mesh (Squamish), xʷməθkwəy̓əm (Musqueam) and səl̓ílwəta (Tsleil-Waututh) peoples; working class settlers (including generations of 'white' East Siders; Chinese newcomers from the mid-nineteenth century onward; the now diasporic Japanese Canadian communities; a substantial African-Canadian community;[11] diverse newcomers from all continents; and the Indigenous peoples who have since found their way to Vancouver for a myriad of reasons.

Interviews with over 50 current and past inhabitants of the DTES/Paueru-gai identified four aspects of the Right to Remain. These are shared in-depth elsewhere (Franks et al. 2015a, NNM 2015), but we briefly summarize them here:

- The *existential* element of the Right to Remain exposes the assaults made possible when state-authorized or condoned oppression targets the socially marginalized and persists over multiple generations.
- The *material* dimensions of the Right to Remain emphasize the right of DTES inhabitants to live – in their neighbourhood – with dignity, safety and an expectation of care and respect.
- The *cultural* Right to Remain declares the right to the social, cultural and spiritual resources that enable DTES/Paueru-gai residents collectively to create livelihoods, spaces of care and support, and opportunities for comfort and pleasure.
- The *political* Right to Remain recognizes the 'double edged' yet intimate relationship between the DTES/Paueru-gai and the rest of the city (indeed the nation), and the rightful role of its inhabitants and allies in its own development.

Of course all four dimensions are political and prescribe the self-directed right to exploit the differences that DTES residents have both created and been marked with by others. Constant themes of the recovery of the DTES story are continuous community and the dialectical relationships between historical presence and absence, and public visibility and invisibility.

The Right to Remain Community Fair

The Right to Remain Community Fair was originally intended to be a collaboration with competitively commissioned artists considering themselves part of the DTES community who would create public art connecting past and present dispossessions. Such a project would have yielded perhaps more typically permanent and identifiable public art, but with a trade-off in decreasing ownership of the creation process for DTES residents. Karen Ward, a member of Vancouver's Gallery Gachet, observes:

> Official Vancouver [is] very keen on public art . . . but it doesn't actually comment in a serious way about what's going on in the city . . . Very little of it comes out of the community where it actually is. And I see that as a problem because . . . I feel that when I walk through my neighbourhood I see public art everywhere. I see it in the marks that people leave on the sidewalk, I see it in the graffiti on the walls, I see it in the way . . . people interact with their own public spaces. . . . And in that sense public art is about claiming space.
> (Franks 2015d)

Embracing the sort of public art Karen describes, we established a less mediated participatory style, with the result that the word 'Fair' began to resonate in different ways; the Right to Remain was not only *for* but *of* the residents. The Fair was directed by a long-time DTES activist, coordinated by an arts facilitator highly respected in the community and facilitated by four artists who work and live there. The approach of the Fair was to be, in the director's words, 'cheap and cheerful': the word 'fair' itself evoking craft and local talent. It would not be a place of trauma and suffering, but one creative and nurturing.

Ultimately, the Right to Remain Community Fair, administered and supported by Gallery Gachet staff and members, engaged approximately 1,100-plus DTES residents as well as Japanese Canadians, among other visitors to the neighbourhood, at over a dozen events (Table 9.2).

The Fair: public art as praxis

The Fair worked as a public art intervention, playing with artistic *form*, contesting and making *space*, and creating encounter and enabling *dialogue*. Academic theorization of this intervention was secondary to ensuring its relevance to participants in the DTES and the Japanese-Canadian diaspora. For the discourse of human rights and the domain of cultural production to be relevant in the highly personal interactions of the Fair, research participants, facilitators and project team members gathered around materialist and experiential understandings of these concepts.

The matter of human rights

In some critical academic and radical political circles, human rights discourse is dismissed as ineffective in revealing the capitalist, colonial relations that formed the DTES. For geographers like David Harvey (2013), however, human rights can be redefined through asserting their socio-spatial characteristics in the city. Iris Marion Young's (2000) more pointed emphasis on inclusion and dialogue as the basis for participatory democracy provides a template for re-envisioning community and challenging oppression. The Fair marked an enthusiastic effort to open up rights-based discourse across multiple engagements with inhabitants and visitors within one rich site of human rights struggle. Unlike urban planning efforts to objectify residents and contain rights (Shoval and Strom 2009), the Fair recognized the DTES as a 'rights-space'. Participating residents were often pleasantly surprised that their experiences/thoughts were valued as contributions to a historically informed narrative of human rights struggles. People expressed a variety of personal expectations of rights (Franks et al. 2015a), including appeals to common humanity ('Why do we deserve that? That's inhumane to do that to anybody'. – 'Audra' (17)) and citizenship status ('[I have a right] to be a citizen, not a lower-class piece of trash'. – Glenn (15)), but also specific rights to bodily integrity ('If we are mak[ing] the decision to sell our bodies, we are making a decision'. – Suzanne (28)), to Indigenous territory ('The state of my rights? A lot better than they were back home. This is unceded Coastal Salish land'. – Michael (33)), and to livelihoods, particularly public street vending ('What's the sense of having a constitutional right when the city can still put up an unconstitutional bylaw that keeps us from vending?' – Tom (15)).

Respondents' lived experience of having/not having rights in the DTES reflected both the economic basis of rights-denying inequities and the persistent, pervasive denial of rights across existential, material, cultural and political realms. As we noted in our final report:

> Many interviewees are keenly aware of the broader economic structures that have shaped and threaten the DTES/Paueru-gai, and attribute these conditions to aggressive capitalism and financialisation of the economy. For long-time Strathcona resident 'Kazuo', a Japanese-Canadian Elder who moved to Vancouver from Japan in the late 1960s, 'commercialism and capitalism . . . are really the core of politics or social involvement today'. 'Wendel', an Elder living in Chinatown, felt that 'everything [is] driven by capital force, so when the prices are going up and you have no choice'. Finally Chanel, a current DTES resident, Indigenous woman and worker at PACE, reflected that 'there just seems to be a deep sense of community down here that's being torn apart by high finance'.
>
> (Franks et al. 2015a: 19)

Our findings support a diverse materialist reading of human rights that connects the chronologically and 'culturally' separated struggles of the racialized and marginalized inhabitants of the DTES.

The matter of culture

In an era when 'culture' and 'creative industries' are prime instruments in creating the 'postpolitical' conditions for erasing agonistic struggles and commodifying urban labour and spaces (Davidson and Iveson 2015), it may seem self-defeating to claim special significance for cultural activity in the DTES/Paueru-gai. But we have sound reasons. The neighbourhood has one of the highest concentrations of practising artists in the city, perhaps the country (City of Vancouver 2014, Walling, Hunter and Greenaway 2015), and houses major cultural industries and arts education. Securing space and investment opportunities for 'creative industries', a heritage/aesthetic/preservation element of neighbourhood revitalization (particularly 'Japan' and 'Chinatown'), and investments in public art are key elements of the City's recent Local Area Planning Process document (City of Vancouver 2014; LAPP).

Beyond these metrics of art and culture, for decades the DTES has been a hub for creative practices of community making, resistance, healing, commemoration and celebration (Franks et al. 2015b, Walling, Hunter and Greenaway 2015). While complexly intertwined with local formal and informal economies, creative practice has deep and immutable 'use value' for DTES residents. This was also true of the Japanese-Canadian population 75 years ago, but their[12] tradition has recently been mined by City of Vancouver planners for its potential transformation into exchange value within the marketized revitalization of the neighbourhood, shifting cultural practices into 'vibrancy' (City of Vancouver 2014, Franks et al. 2015b) and selective working-class and ethnic histories into 'heritage'. Reflecting on her experiences in a poorly maintained 'heritage' building, Karen Ward said 'this idea of the heritage building [designation] was this thing that the landlord was trying to get – I mean basically they want to increase the value of the space, and that was their only hook really'. Residents also understand that the profit-enhancing (rent-increasing) potential of a 'heritage' designation is directly tied to the broader homogenizing project of urban regeneration, as 'Richard', DTES resident, activist, and videographer, encapsulates:

> The city's gotta be careful . . . it is such a mixed community . . . people come from all kinds of backgrounds, and to me that is the essence of the community, so if you want to sterilize it and say this is just gonna be Japantown, no, you know, you can't bring the past back to the present.
>
> (Franks et al. 2015a: 23)

No one is actually promoting a demographic *return* to 'Japantown' in the Powell Street/Paueru-gai area. But what is increasingly mooted is a selective application of easily consumable cultural products that together manifest a heritage brand of 'Japantown' (Masuda and Bookman forthcoming). Not only is this brand superficial for a largely working-class and hybrid Japanese-Canadian culture, its ahistoricity is easily spliced into other branding exercises, giving the Downtown Eastside the complexion of a micro-branded residential theme park. For example, the high-end restaurant Cuchillo on Powell Street locates itself on Twitter in 'JapaGasRailTown', a meaningless fusion of the adjacent Japantown, Gastown and Railtown districts. Even the area's Powell Street Festival is a contested site – born of grassroots Japanese-Canadian activism but also functioning as an ephemeral spectacle that can transplant historically entwined urban relations into Vancouver's consumption-oriented 'event landscape' (Kroeber 2015).

The Fair in form, spatiality, and dialogue

Artistic form, spatiality and dialogue are aesthetically and politically connected in many ways. What makes the Fair unique is the near-complete autonomy the art team had within the limits of finances and time, enabling diverse and spontaneous connections between aesthetic form and spatial engagements, furthering capacity for dialogue within and between members of low-income, Indigenous, and Japanese-Canadian communities either inhabiting or allied to the DTES by politics and/or ancestry. Team members were selected for their skills and experience, but also their community status as good neighbours and committed advocates. Once established, the team collectively decided that each workshop would be led by one artist who would choose the place and the media (Table 9.2), with the rest of the team in supporting roles.

Table 9.2 The Right to Remain Community Fair. July 2014–January 2016

Event type and activity	*Place and Date*	*People*
Workshop: storyboards with Andy Mori	Aboriginal Front Door (AFD), 17 July 2014	RRCF team; DTES residents
Workshop: dioramas with Karen Ward	VANDU (Vancouver Area Network of Drug Users), 24 July 2014	RRCF team; DTES residents
Powell Street Festival (PSF); Tent City 'occupation': Postcard exchange project	Oppenheimer Park; adjacent streets, 1–3 August 2014	RRCF team; DTES and tent city residents; Japanese Canadians; PSF attendees
Workshop: site photos and reflections with Quin Martins	Interurban Gallery, 21 August 2014	RRCF team; DTES residents
Workshop: Image theatre with Herb Varley	Gallery Gachet, 18 September 2014	RRCF team; DTES residents
Workshop and museum tour: Postcard exchange and talking circle; museum tour	Nikkei National Museum, Burnaby, BC; 8 October 2014	RRCF team; DTES residents; Japanese Canadians; Museum staff
Single Room Occupancy (SRO) Convention: 'Making the Links' Art Cart workshop	Vancouver Japanese Language School and Japan Hall (VJLS-JH), 19 October 2014	RRCF team; DTES SRO residents; housing activists; VJLS-JH staff
Heart of the City Festival: Public Presentation; artist talks	Gallery Gachet, 8 November 2014	RRCF team; Japanese Canadians; Festival and gallery visitors
Walk. Eat. Talk. Make: Walking discussion; group meal; *ema*	VJLS-JH; Carnegie Community Centre, 10 January 2015	RRCF team; DTES residents; Japanese Canadians; community advocates
The Right to Remain: Gallery exhibit; talks; interactive stations	Gallery Gachet, 6 March– 12 April 2015	RRCF team; Gallery Gachet members; general public
Exchanges: Presentations; open public talking circle	Gallery Gachet, 12 April 2015	RRCF team; DTES residents; Japanese Canadians; general public
Labour Landscapes: a Storytelling Walk: Walking group; storytelling; presentations	Powell Street Festival and Centre A, 1 August 2015	RRCF team; DTES residents; Japanese Canadians; general public
Revitalizing 'Japantown'?: a *Right to Remain* Exhibit: Gallery exhibit; talks; interactive stations	Nikkei National Museum, 24 October 2015–31 January 2016	RRCF team; general public; Japanese Canadians; DTES residents

With one team member taking responsibility for explaining and demonstrating the art-making on offer, other team members could assist participants and one another in a variety of ways. Each Fair event (Table 9.2) also featured a casual 'come and go'

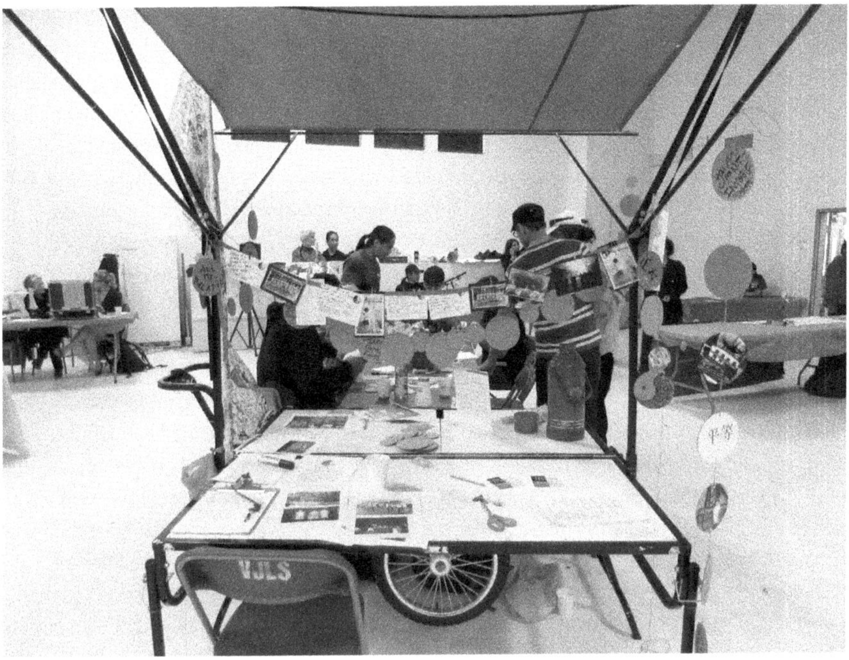

Figure 9.1 Banner making at The Art Cart outside a workshop venue. Source: Ali Lohan.

element. Everywhere the Fair went, the Art Cart went,[13] and scores of neighbours and passers-by paused to make buttons, add to an evolving banner, and talk about the Right to Remain. At least one team member stayed outside the workshop venue with the Art Cart, answering questions and inviting people inside (Figure 9.1).

Each event required art material to be set up (and cleaned up), and tables and chairs, food and drink to be arranged. Care was taken to make sure that those who did not wish to be photographed or filmed would not be. And because this was a research process, the team, led by research assistant Trevor Wideman, ensured that everyone participating was aware their work was on record and consented in writing to having it photographed and written about.

In the following sections, we address how artistic form, spatiality and dialogue played out in the choices the art teams made, the serendipitous encounters that resulted and how materials and space were engaged to engender dialogical encounters – conversations that revealed and re-historicized previously effaced continuities between different communities and their dispossession through colonial capital.

Form, or format

A quick scan of the 'Activity' column in Table 9.2 illustrates the breadth of artistic media at the Fair: plastic arts (dioramas), image-based work (photo journaling), embodied performance (image theatre), neighbourhood walking, discussion and storytelling in the vein of 'social practice' art (Bishop 2012, Kester 2013), as well as forms that spanned many media as processes (making), objects (lasting 'things'), and events (acts of meeting/exchange), in particular the recurring use of postcards (Franks et al. 2015b). In keeping with our ethos of resident-artist-led interventions, we offer the artists' own analysis of the particular impacts of their formats, focusing on two: storyboards and dioramas. Both are small scale but invoke different temporalities, relationships to place and affective responses.

Storyboards were facilitated by Andy Mori at the Aboriginal Front Door and Indigenous drop-in centre. A storyboard is a simple set of panels, like a comic book, where filmmakers visualize their stories with pictures and drawings. Andy's storyboard templates featured sheets of paper with pre-printed panels: 'So there was a blank rectangle on the page for them to fill in with a picture, and there'd be some lines underneath the frame for their scribbled notes about what was happening in the scene' (Franks 2015b). This format allowed simplicity and ease of use for both participants and his fellow team members: 'If [participants] wanted to do stick people I said okay . . . I was trying to find the lowest barrier format . . . I could for drawing . . . for the team, [they] did not have to worry about me so much. They can worry about their own roles' (2015). The 'low barrier' format also encouraged creating a safe space to work through rights issues, inviting powerful biographical reflections through bare-bones narratives. Andy's second reason for using storyboards proved deeply emotional for participants, as he told workshop participants:

> It's like a movie that we're not going to film . . . you're going to write an internal monologue, good or bad, about a human rights issue that you personally encountered in the DTES . . . this is like if you're storyboarding . . . a section of your life about a human rights event in the DTES, what would it look like?
>
> (Franks 2015b)

In Andy's view the storyboard provides a format to explore longer life histories in a condensed fashion – thumbnail auto-ethnographic accounts of lived experience with human rights struggles that don't require extensive writing or drawing skills. Because they needed to 'get to the point' in only six frames, these life histories were also explicitly evocative of place: 'there was a sense of home. This was their stories in their place in their context . . . and there was this longing for security, which is another part of the [workshop] process' (Franks 2015b).

Diorama making – a 'little kind of tableaux or scenes or constructions within a frame, within a box' (Franks 2015d) – was facilitated by Karen Ward at the Vancouver Area Network of Drug Users (VANDU), a drop-in centre and advocacy organization. Like the storyboard workshop, the objects created were small (paper boxes about 10 inches square), and both the six-panel template sheets and the paper boxes (takeaway containers donated by project partner Potluck Café) provided ready-made platforms upon which participants' human rights experiences could be enacted. The diorama format utilized a wide range of materials, including small plastic figures, photos, scrap paper, found 'odds and ends' and fabric. Karen explained the selection of materials (Figure 9.2):

> So we've got members of VANDU [at the workshop], and we're going to have members from WAHRS, which is Western Aboriginal Harm Reduction Society, so . . . I thought 'I'm going to go to the toy store and buy plastic figures of quote-unquote "Indians" and "soldiers"' . . . basically to amplify conflict in the kind of the 'colonized' and the 'colonizer' broad narrative in the neighbourhood.
>
> (Franks 2015d)

In addition to these colonial tropes, Karen offered two-dimensional images of the rapidly gentrifying urban environment for use in the diorama space:

> I took photographs in the neighbourhood . . . pictures of signs . . . things on the sidewalk, of the teepee that was set up in Oppenheimer Park . . . of the barriers that are set up by development and revitalization, the kind of 're-branding' that had been going on . . . it's a way of marketing the DTES. . . open for further development. . . . I took pictures of ... street signs that say 'No Camping, Bylaw, fine imposed' . . . I cut them to fit the boxes in various ways that people wanted to use them, as collage or background or whatever.
>
> (Franks 2015d)

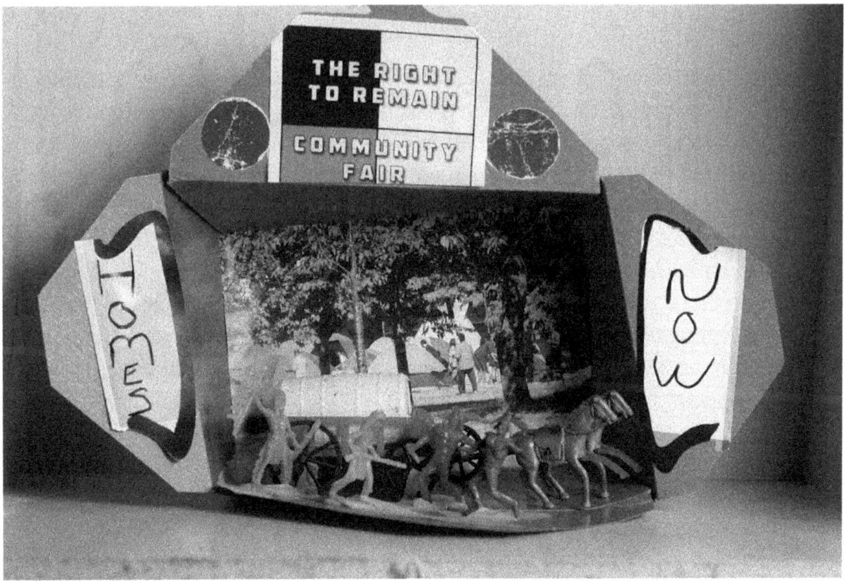

Figure 9.2 Diorama: *Homes Now* by DJ Joe. Source: Trevor Wideman.

By landscaping the paper boxes with materials evocative of spatial dispossession, participants made three-dimensional claims upon the DTES, crafting it as a place dispossessed historically through colonization and now through market-based gentrification. The storyboard and diorama formats were diverse ways of engaging participants (drawing, writing, sculpting, crafting) and intervening in public DTES space-times.

Making space

All aesthetic forms play with spatiality; because the Fair was interactive, multi-sited and operated through long-standing social networks in the DTES, its spatial dimensions transcended specific art forms and individual art-making (Table 9.2). The Fair operated on the DTES streets; in intimate gatherings (the workshops at VANDU, AFD and Gallery Gachet); outdoors on the 'concrete street' (nodding to Lefebvre (1992), the DTES built environment, and Freire's notion of the 'concrete' in popular education practice (1996)); and through institutional spaces (particularly the venues of the project partners, such as the Powell Street Festival and Gallery Gachet).

At each venue the team strove to create a focused but open, self-regulating atmosphere. Karen describes the scene at VANDU:

> Just having that ability to feel able and free to say, 'Yeah, this is what I'm doing, yeah this happened to me, this happened to my grandmother,' and having that safe space in which to both talk and share and make at the same time. . . . in a very broad sense it was all collaborative.
>
> (Franks 2015d)

An easily overlooked element of this collaborative spatiality is the interior 'spaces', or spaces of change and self-actualization, within participating subjects. Herb Varley's workshop at Gallery Gachet used variants of Augusto Boal's Image Theatre (1995) to explore personal experiences of gentrification, violence and marginalization in the public spaces of the DTES. For anyone unfamiliar with applied theatre, this form may seem to involve the least individual introspection, given its basis in reception and collective creation. In Herb's view, however, both the creators and witnesses of intense imagery in Image Theatre (who alternate roles throughout) generate deeply personal affects that can lead individuals to challenge the stereotypical social roles – and the urban spatial designations – imposed on them by others:

> [I]t allows you to really explore those feelings and what kind of emotions well up . . . for the audience, it's very visual, but for the people making the tableaux it's very physical . . . it's unusual to get a group of strangers to explore the issues of gentrification and how it dehumanize[s], or the minimizing of human rights. . . . Part of what we're trying to do with the Right to Remain is . . . we're artists. We think. We're political beings, we're spiritual beings . . . coming from an Aboriginal perspective, and also . . . Carl Jung's work and

> the process of 'individuation.' . . . I like that idea of getting people to focus on their growth as human beings.
>
> (Franks 2015c)

Such intimate styles of creative work and spatiality were timed intermittently throughout the Fair with events more clearly engaged with 'the concrete street'. Quin Martin's workshop based at the open storefront of the Urban Gallery on East Hastings Street invited participants to go out and photograph 'areas that are meaningful to them in some ways' (Franks 2015a) with Quin's cellphone, then email the picture back to a laptop, where an assistant would transcribe the participants' personal commentary on the significance of that place. Quin explained that the cell photography medium has two main aims: 'salvaging' or reclaiming the class character of the neighbourhood, and using photographic practice to reclaim the agency and subjectivity of low-income residents who are far more often objects in photos than photographers themselves:

> [W]e're living in sort of this DTES that's a disappearing culture and so, we're salvaging some kind of visual document of the neighbourhood. Because it's not gonna be there much longer. . . . I wanted people to include themselves in the photograph because often in the neighbourhood, people are photographed without their consent and so, rather than people going out in the neighbourhood and photographing other people, I wanted them to include themself, as a subject.
>
> (Franks 2015a)

The Fair also intervened on the 'concrete street' in the way it artfully, at times delicately, translated rights discourse and lived experience between the low-income (particularly homeless) and Japanese-Canadian communities. Based on an idea by Andy Mori, the Fair team devised an exchange of specially designed postcards (Franks et al. 2015b) between visitors to the 2015 Powell Street Festival (PSF) and homeless people and their allies, many Indigenous, who had in the weeks leading up to that year's festival created a small 'tent city' or protest encampment in Oppenheimer Park in the heart of the DTES to draw attention to homelessness and the City's disregard for Coast Salish territorial rights. For 37 years the PSF had been held in the Park, a vital social and cultural space for the pre-1940s Japanese-Canadian community that referred to the site as Powell Ground. In 2015, out of solidarity with the homeless camp, the Festival was moved onto adjacent streets and sidewalks. In facilitating the postcard exchange, Andy and Karen describe how the team navigated two very different spaces that delineated a semi-public 'grey zone' of belonging and identity.

> It was a mixed response at first . . . talk about different cultures right? So we have a festival crowd who is almost like 'catch and release' – we catch them for a minute then we release them. But, tent city was more hermetic, like they were in it and they were protecting it, and so [we had] to send an envoy, like Herb and Karen, just sort of get in there and sort of explain what [the project] was.
>
> (Franks 2015b)

Having acted as an 'envoy' between groups merely a block apart, Karen describes their experience:

> When I was [at the Camp] it was mostly a group of women and we were just talking about how great it was that other people . . . hadn't forgotten about the people in tent city and that there was support outside of the neighbourhood . . . And then . . . Herb was telling the story of the [Japanese-Canadian] internment to people in the camp, and most people didn't know anything about it. And so . . . there were conversations about land and history and language and all these things. And again, about continuity in the neighbourhood and about recognizing the parallels, the obvious cycles of displacement. So I think . . . we were literally drawing a line between the tent city and Powell Street Festival that maybe wasn't as obvious as it might have been.
>
> (Franks 2015d)

Finally, we turn to these acts of 'drawing a line between'.

Dialogues

A fundamental aim of the Fair was to use human rights discourse to catalyse an emergent dialogue between current, low-income, often racialized inhabitants of the DTES and members of the Japanese-Canadian community, some of whom are descendants of families forcibly driven from the neighbourhood during the 1940s. Because of the demographic and geographical distance between these groups, our methods needed to be much more directive.

On 8 October 2014, project partner and Fair contributor the Nikkei National Museum (NNM) hosted a workshop and museum tour at their gallery in suburban Burnaby, BC, a trip that is 45 minutes by public transit from the DTES, but much more distant socio-economically. Including facilitators from the DTES and NNM members, seven residents of the DTES travelled as a group and met with ten Japanese Canadians, several of them quite elderly. After artists Andy Mori (of Japanese-Canadian descent) and Herb Varley (of Nisga'a and Nuu-Chah-Nulth descent) introduced the project, the group shared tea and conversation while adding to the banner and writing postcards. Project contributor Kathy Shimizu, who began working with the Fair during this shift of focus to DTES–Japanese-Canadian dialogue, comments that these activities 'provide a way for people who do not usually meet face-to-face . . . to increase understanding between people from different backgrounds, and build connections between communities' (email communication 2015).

But she emphasizes that the real artistry of the event was not in the art forms but in exchanges 'which were not staged but happened spontaneously, [and] were quite profound' (2015). For Kathy these included the gifting of a handmade rattle by two Indigenous Elders and long-time homelessness activists from the DTES to the eldest Japanese-Canadian person in the room. Kathy recalls another Japanese-Canadian elder expressing surprise about the negative connotations of the word 'gentrification', assuming it meant positive investments by residents in improving their own community, much as the Japanese Canadians of Paueru-gai had done decades before. As she recalled, 'The clarification between improvements in a neighbourhood for those that live there and improvements that come in that result in sweeping people out was a critical point of the workshop, and helped bring a different perspective to the discussion.' While modest in numbers, this event (and others like it) specifically intended to create dialogue between the low-income DTES and Japanese-Canadian communities (Table 9.2) and provided what would be for many the only opportunity to meet face-to-face, 'showing their eloquence and intelligence to those who might judge them differently' (2015). The Fair embodied a restorative narration of a contentious but continuous community, where the very act of (re)introducing marginalized groups to one another foregrounds the shared abuses of their human rights, the Right to Remain in particular.

Conclusion

The DTES is a place where diversity is a result of historical layers of residence by Indigenous peoples, working-class Chinese and Japanese immigrants and their descendants, and the accretion of poverty as a result of the inaccessibility of civil rights, employment, health care and affordable housing to the people who live in the neighbourhood today. Our larger goal, allied with those who live there, is to stop the dispossession. The Fair team used assertive, communicative interaction in the neighbourhood's public and semi-public spaces as palpable (if not quite tangible) interventionist public art. The project mounted resistance to the enclosure and commodification of public places through autonomous but collective manipulation of form, spatiality and dialogue. We have sought to convey how this project, through human dialogue and creative art forms, addressed the ongoing colonial erasure of First Peoples upon which the supposed revitalization of the DTES is predicated. If these 'successes' were partial and incomplete, their strength lies in how they emerged from existing social relationships and creative practice in the DTES. We also suggest that these relationships and practices continue (for the moment) to defy enrolment as 'exchange value' in an increasingly hyper-marketized urban landscape, maintaining significant 'use value' for residents as a mark of DTES distinction. Finally, while spaces of dialogue were crucial features of all Fair events, we need to focus more attention on engineering links across different historical and contemporary constituencies in the DTES/Paueru-gai to engage the Japanese-Canadian diaspora, a key linking agent between past and present in the ongoing colonization of urban Vancouver. Resisting gentrification and defending diversity in the DTES means incorporating a dialectical relationship between past and present communities, strategic visibilities and invisibilities, inalienable rights to security and difference, and perhaps most crucially, grounded claims to territory and the dispossession of Indigenous, marginalized and racialized communities.

Acknowledgements

We extend great thanks to the Right to Remain Community Fair team of artist-facilitators, and to our formal and informal partners in the DTES and Japanese-Canadian communities. We thank Savannah Walling for identifying a wealth of written resources on the history of art-making in the DTES. We are grateful for the continued presence of Indigenous peoples in the DTES, including the Sk̲wx̱wú7mesh (Squamish), xʷməθkwəy̓əm (Musqueam) and səl̓ílwəta (Tsleil-Waututh) people on whose unceded territory Vancouver now stands. Revitalizing 'Japantown'? was financially supported by the Social Sciences and Humanities Research Council of Canada, the City of Vancouver Community Arts Fund, the University of Manitoba, Queen's University, and the Canada Research Chairs program.

Notes

1 The Right to Remain Community Fair Team are coordinator Ali Lohan; community peer arts facilitators Quin Martins, Andy Mori, Herb Varley and Karen Ward; research assistant Trevor Wideman; local activist gallery Gallery Gachet; and the National Nikkei Museum.

2 Paueru-gai is a Japanese iteration of 'Powell Street', the main thoroughfare that runs through the neighbourhood.

3 Named after the British sea captain who charted the waters.

4 'Skid Road' refers to the log skids laid down to transport logs (from a cutting site to water where they were transported in large booms), and along which workers' shacks were erected.

5 These include earlier incarnations of the Japanese Canadian Citizens Association (Greater Vancouver Chapter), the Japanese Canadian Society of Greater Vancouver, and *Tonari Gumi* Japanese Community Volunteers' Association.

6 The research team conducted interviews with over 50 current and past DTES residents, including Japanese-Canadian and Chinese-Canadian elders and Aboriginal people. These interviews are the backbone of the project's community resource, 'The Continuous Fabric of Change, Resistance and the Right to Remain in the DTES/Paueru-gai' (Franks et al. 2015a). Participants who wished to be anonymous have been given pseudonyms indicated by single quotation marks.

7 This does not include the 50-plus interview participants, the 40-plus community residents and allies who attended Advisory Committee meetings, or the hundreds of DTES inhabitants and allies who participated in the Right to Remain Community Fair activities.

8 The roles listed here are not exhaustive.

9 The SBIA ceased participating in the project in 2013.

10 While an exhaustive consideration of this concept is beyond our scope, addressing superdiversity in this context raises the question of whether it is indeed blind to the particular colonial conditions within Settler societies.

11 For information on the dispossession and dispersal of the Black community, see http://blackstrathcona.com/about/; and Kilian 2008.

12 Some Japanese-Canadian elders indicated that one of the long-term consequences of the uprooting (1942–1949) was that, due to the enormous disruption to community life and people's livelihoods, art and culture were neglected as expendable luxuries (Franks et al. 2015a).

13 The Art Cart is a portable, wheeled art-making station with built-in table-top and storage. It is maintained by Gallery Gachet and is a highly visible and welcome presence in the DTES.

References

Adachi, K. 1976. *The Enemy That Never Was*. Toronto: McClelland and Stewart.

Berg, M. and N. Sigona. 2013. Ethnography, diversity and urban space. *Identities*, 20: 347–360.

Bishop, C. 2012. *Artificial Hells: Participatory Art and the Politics of Spectatorship*. London and New York: Verso.

Boal, A. 1995. *The Rainbow of Desire*. London: Routledge.

Centre for Environmental Health Equity (CEHE). 2011. *Where Everyone Is Sort Of, 'Me and Mine': Relocating Environmental Health Equity in the City through the Photos of Inner City Residents in Vancouver, Winnipeg, and Toronto* [Online pdf]. University of Manitoba, Winnipeg, MB. Available at: www.cehe.ca/sites/default/files/SUCCEED%20FINAL%20REPORT.10Mar2011.pdf [accessed: 21 November 2015].

Christoff, S. and Kalache, S. 2007. The poorest postal code: Vancouver's Downtown Eastside in photos. *The Dominion: News from the Grassroots*, 42. Available at: www.dominionpaper.ca/articles/909 [accessed: 17 December 2015].

City of Vancouver. 2014. Downtown Eastside Plan [Online]. Available at: http://vancouver.ca/home-property-development/dtes-local-area-plan.aspx. [accessed: 24 November 2015].

Cresswell, T. 2006. *On the Move: Mobility in the Modern Western World*. New York: Routledge.

Davidson, M. and Iveson, K. 2015. Recovering the politics of the city: from the 'post-political city' to a 'method of equality' for critical urban geography. *Progress in Human Geography*, 39(5): 543–559.

Franks, A. 2015a. Interview with Quin Martins. Interviewed by Aaron Franks. Digital recording. Powell Street Festival offices. 27 March 2015.

Franks, A. 2015b. Interview with Andy Mori. Interviewed by Aaron Franks. Digital recording. Powell Street Festival offices. 27 March 2015.

Franks, A. 2015c. Interview with Herb Varley. Interviewed by Aaron Franks. Digital recording. Gallery Gachet. 28 March 2015.

Franks, A. 2015d. Interview with Karen Ward. Interviewed by Aaron Franks. Digital recording. Gallery Gachet. 28 March 2015.

Franks, A., Masuda, J. and the Revitalizing 'Japantown'? Research Team. 2015a. *The Continuous Fabric of Change, Resistance and the Right to Remain in the DTES/* Paueru gai*: a community research resource*. Kingston, ON: Queen's University, DOI 10.13140/RG.2.1.1936.1124.

Franks, A., Mori, A., Lohan, A., Masuda, J. and the Right to Remain Community Fair Team. 2015b. The Right to Remain: reading and resisting dispossession in Vancouver's Downtown Eastside with participatory art-making. [Online *Feral Feminisms* 4]. Available at: http://feralfeminisms.com/right-to-remain/ [accessed: 21 December 2015].

Freire, P. 1996. *Pedagogy of the Oppressed*. London and New York: Penguin.

Globe and Mail. 2009. Our nation's slum: time to fix it. 14 February.

Harvey, D. 2013. *Rebel Cities: From the Right to the City to the Urban Revolution*. London: Verso.

Hopper, T. 2014. Vancouver's 'gulag': Canada's poorest neighbourhood refuses to get better despite $1M a day in social spending. *National Post*, 14 November.

Humphris, R. 2015. Intersectionality and superdiversity: what's the difference? [Online]. IRiS Key Concepts roundtable series, Institute for Research into Superdiversity. Available at: www.migrationsystems.org/intersectionality-and-superdiversity-whats-the-difference/ [accessed: 24 November 2015].

Kester, G. 2013. *Conversation Pieces: Community and Communication in Modern Art*. Berkeley, CA: University of California Press.
Kilian, C. 2008. *Go Do Some Great Thing: The Black Pioneers of British Columbia*. Burnaby, BC: Commodore Books.
Kroeber, G. 2015. The event landscape: the contemporary encounter of art and the city. *PAJ: A Journal of Performance and Art*, 37: 30–41.
Lefebvre, H. 1992. *The Production of Space*. Malden, MA and Cambridge: Wiley-Blackwell.
Malone, P. 1941. City on war basis as Pacific erupts. *The Vancouver Sun*. 8 December.
Massey, D. 2007. *World City*. Cambridge, UK, and Malden, MA: Polity.
Masuda, J. and Bookman, S. Forthcoming. Neighbourhood branding and the right to the city. *Progress in Human Geography*.
Masuda, J. and Crabtree, A. 2010. Environmental justice in the therapeutic inner city. *Health and Place*, 16(4): 656–665.
Masuda, J., Zupancic, T., Crighton, E., Muhajarine, N. and Phipps, E. 2014. Equity-focused knowledge translation: a framework for 'reasonable action' on health inequities. *International Journal of Public Health*, 59(3): 457–464.
Mickleburgh, R. 2013. Buildings to soar over Chinatown after Vancouver eases height restrictions. *Globe and Mail*. 24 February.
Miki, R. 2004. *Redress: Inside the Japanese Canadian Call for Justice*. Vancouver: Raincoast Books.
Nikkei National Museum (NNM). 2015. Revitalizing 'Japantown'? A Unifying Exploration of Human Rights, Branding, and Place [Exhibition catalogue]. Burnaby, BC: Nikkei National Museum and Cultural Centre.
Robinson, D. and Martin, K. 2016. Introduction: 'bis a resonant chamber'. In *Arts of Engagement: Taking Aesthetic Action in and beyond Canada's Truth and Reconciliation Commission*. Waterloo, ON: Wilfrid Laurier University Press.
Shimizu, K. 2015. Email from Kathy Shimizu. RRCF event report to Aaron Franks. 12 April 2015.
Shoval, N. and Strom, E. 2009. Inscribing Universal Values into the Urban Landscape: New York, Jerusalem, and Winnipeg as Case Studies. *Urban Geography*, 30(2): 143–161. DOI: 10.2747/0272-3638.30.2.143.
Takeuchi, K. 1982. Japantown beautification. [Letter to mayor]. Box: 54-G-3 folder 9. Vancouver: City of Vancouver Archives.
Teelucksingh, C. and Masuda, J. 2014. Urban environmental justice through the camera: understanding the politics of space and the right to the city. *Local Environment*, 19: 300–317.
Vertovec, S. 2007. Super-diversity and its implications. *Ethnic and Racial Studies*, 30: 1024–1054.
Walling, S., Hunter, T. and Greenaway, E. 2015 *From the Heart of a City: Community Engaged Theatre and Music Productions from Vancouver's Downtown Eastside, 2002–2013*. Vancouver: Vancouver Moving Theatre.
Ward, P. 1978. *White Canada Forever: Popular Attitudes and Public Policy toward Orientals in British Columbia*. Montreal and Kingston: McGill-Queen's University Press.
Young, I. 2000. *Inclusion and Democracy*. Oxford and New York: Oxford.

10 The art of (re)crossing the border

The *Border Farm* project in Maroi, South Africa

Pauline Guinard

Introduction

Surely less well known than other borders, the South Africa/Zimbabwe border is nevertheless one of the busiest in (Southern) Africa. First established along the Limpopo River in 1881, that is to say after the first Boer war,[1] the South Africa/Zimbabwe border has since been stable. Two hundred and twenty-five kilometres long, it goes from the western tripoint with Botswana, to the eastern tripoint with Mozambique. Despite its length, there is only one gate and one bridge, both named Beitbridge, to cross it (Figure 10.1). Consequently, the South Africa/Zimbabwe border is very active and super-diverse. According to a South African border employee, there are about 30,000 persons crossing every eight hours, and even more at month ends, during Easter and Summer holidays. People crossing come from Zimbabwe, Zambia, Malawi, Tanzania, Somalia, etc. Due to the lack of alternative routes and the proximity of the economic hub of Johannesburg, Beitbridge border post acts as a bottleneck where people from all over Southern and Eastern Africa converge to enter South Africa. Besides these legal crossings, this area is also famous as a space for illegal ones. Every day people are crossing through the Limpopo River, risking being robbed or assaulted by smugglers who take advantage of the border jumpers, when they do not drown in the river. Characteristic of an attempt made by various states around the world to control and secure their borders – especially against illegal migration and migrants both seen as undesirable (Amilhat-Szary 2015, Amilhat-Szary and Giraut 2015, Andreas 2003, Didiot 2013, Rosière and Jones 2012) – fences and barbed wires have been put in place along the river by the South African state, while the army is patrolling the area. Despite these measures, the holes in the fences and the open gates (as well as, in some places, the absence of the fence) reveal the porosity of this border.

Emblematic of an increasing tension between openness and closure of borders in a context of globalisation (Newman and Paasi 1998, Nicol and Minghi 2005, Wilson and Donnan 2012), the South Africa/Zimbabwe border area is a contradictory space. Its contradictory nature is even more obvious in times of crisis as it was the case in the late 2000s. After the contested re-election of President Mugabe in March 2008, Zimbabwe experienced galloping inflation, political

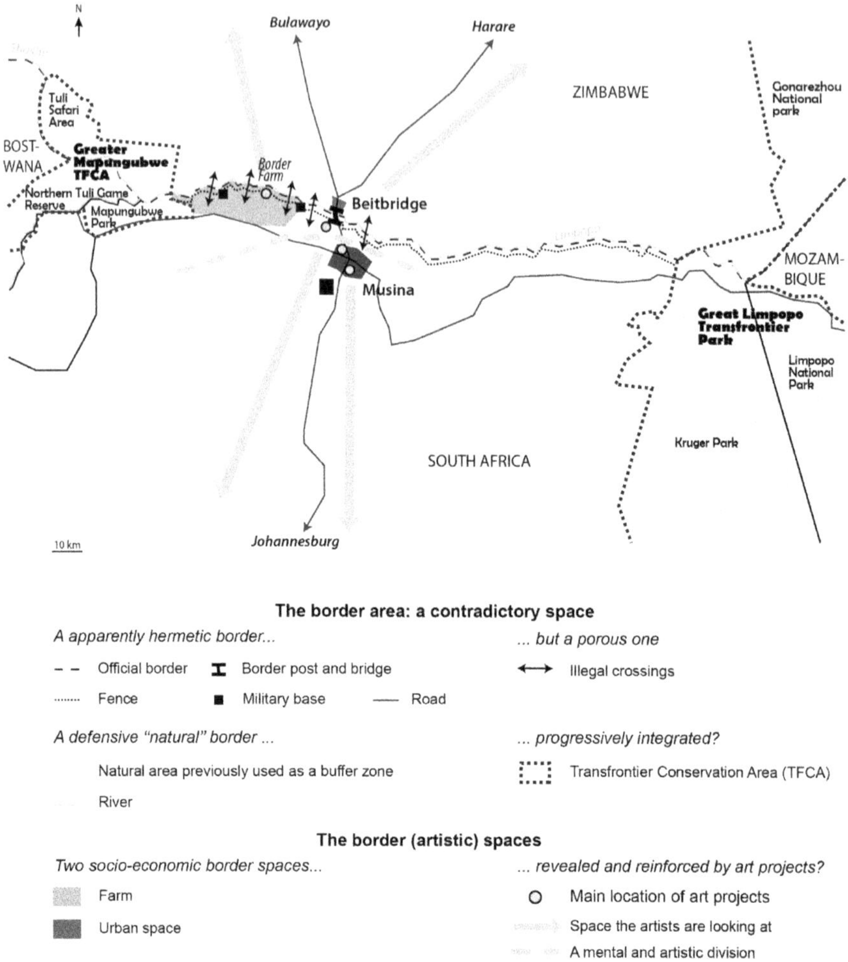

Figure 10.1 The border area, a contradictory space. Source: Pauline Guinard, 2016.

violence and a health crisis due to an outbreak of cholera, forcing thousands if not millions of people to flee the country. These events intensified the tensions previously existing among the different groups of population – especially defined by their nationalities – in the region, while attracting NGOs, journalists and artists. Intrigued if not alarmed by the migrants' situation, several artists actually decided to understand and document the lives of these people in the border region. However, the presence of artists in such a context can at first appear as surprising. Why are artists drawn by these areas? What might be the role of art and artists in this context? As shown by Amilhat-Szary (2012, 2015), the presence of artists in border areas – be they natives of those areas or not

(Berelowitz 1997) – is far from being exceptional nowadays. On the contrary, their importance and their potential capacity to propose alternative border narratives to the ones often shaped by the media are such that this kind of art tends to be recognised as a specific artistic form known as 'border art'. Nevertheless, despite the potentialities of that art in questioning if not contesting the existence of the borders, its effects on them are not unequivocal (Amilhat-Szary 2015, Sheren 2015): it can also contribute to their production and reinforcement by giving them a certain reality and visibility. The role of art and artists in the (un)making of the borders, in the process of re/de-bordering (Amilhat-Szary and Giraut 2015, Janczak 2011), is thus ambivalent and has to be critically explored, especially in a context of super-diversity. As underlined by Vertovec (2007), the increasing complexity of the migration process around the world requires an approach that takes into account not only migrants' nationalities or ethnicities, but also their ages, genders, classes and – even more importantly – immigration statuses. Because immigration status can influence the way one can cross a border, settle in a country or interact with local populations, it has to be acknowledged by the researchers examining the relationships people have with the border, as well as by the artists who claim to do art that is relevant to the people and spaces they intervene in.

To investigate the consequences of the presence of border art and artists in the South Africa/Zimbabwe area, I will look at different art projects that took place in the area during the last decade, and I will pay particular attention to the *Border Farm* one. Initiated in 2009 by Thenjiwe Nkosi, a South African artist living in Johannesburg, this participative film project was based in Maroi on a farm situated along the border that is mainly employing Zimbabwean workers (Figure 10.1). Through the re-enactment of the migrants' crossing by the farm workers, be they (il)legal migrants or not, this film explicitly aimed to bring together the places and people who are divided in the migration process. By studying this project, which is as much a border art project as a public art one, the issue is to understand to what extent art and artists can reveal and act on the socio-spatial divisions at work in a border space in order to build bridges between divided places and people that are super-diverse. The choice of that case study is justified by a will to contribute to the current debate on the spatial, symbolical, social and political effects of public (border) art at an empirical, theoretical and methodological level. Empirically, I aim to enrich the discussion by examining a project in an area away from the famous border sites and large cities, which are often over-represented in literature. Theoretically, I wish to look more closely at three dimensions that are often understudied with regard to public (border) art, namely its impacts: (1) on the so-called 'publics' it is designed for or with (Hall 2003, Zebracki 2011); (2) at various scales and; (3) in the long term. Methodologically, I intend to explore art both as an object and a methodology in order to understand the production and representation of (border) space. Based on observations and interviews I conducted in 2013 and 2014 in the area, I will argue that a public border art project like *Border Farm* can be an opportunity to address migrants' traumatic experiences of (il)legal migration, to try to raise awareness around social inclusionary/exclusionary geographies of migration and, by doing so, to bring people and places closer along their manifold social and cultural differences. In that perspective, I will first look at the capacity of the South Africa/ Zimbabwe border to stimulate artists' imagination and attract them in the area to create projects. Second, I will examine the *Border Farm* project more closely to demonstrate its therapeutic function regarding the individuals and the group. Third, I will wonder about the capacity of art as a geographical tool to apprehend the spatiality of borders.

The South Africa/Zimbabwe border as a catalyst for art

As underlined by Stevens (2007), borders, boundaries and barriers are not necessarily limits or discontinuities: they are also places of potentialities and opportunities that can act as 'catalyst[s] for creative or transgressive behaviour where people test the effectiveness of physical controls' (2007: 114). In that regard, the number of art projects that have been taking place in and around the South Africa/Zimbabwe border since the mid-2000s (Table 10.1) underlines the capacity of this specific border to stimulate artists' creativity and of being a catalyst for art.

All these artistic projects were initiated by South African artists who were not necessarily natives from or based in the border area. Most of them were indeed located in Johannesburg which is the main South African artistic hub. This illustrates the attractiveness of the border for artists, at least on a regional scale. While the United States/Mexico border or the Israel/Palestine one appear as artistic hotspots worldwide (Amilhat-Szary 2012, Lehec 2014), the South Africa/Zimbabwe border could be considered as a *regional* one.

Furthermore, the type of art prompted by the existence of this border seems to be specific. Mainly made of visual art projects, it aims to question the physical presence of the border or its consequences on the social space, in terms of (il)legal crossing, prostitution or (in)formal trade, for example. It is not only conducted in the border region considered both socially and spatially, but it is also produced *with* it and *for* the people living in this area. This form of participative art is symptomatic of different trends that characterise contemporary (public) art. First, it has to be understood as part of what is often referred to as a 'spatial turn' in contemporary art, that was notably initiated by Land artists (Volvey 2007). This refers to the fact that artists are more and more considering space as a material, rather than as a mere container or receptacle of their work. In that perspective, art is made *with* space and not *in* space. This radically transforms the ways art itself is produced. To be more attentive to the places they are in along with the people living in those spaces, artists are developing practices that intend to respond to their socio-spatial environments. These practices are actually quite similar to the ones used by geographers, especially due to the importance given to fieldwork (Volvey 2014) and mapping (Cosgrove 2008). Consequently, they can even nourish the geographical knowledge in return (Hawkins 2011). Both in its material and immaterial dimension, space – and in that case border space – can (in)form the artistic practices and reciprocally. Second, because border art often includes a social and political agenda, such as raising awareness about migrants' situations

Table 10.1 The main artistic projects prompted by the presence of the South Africa/Zimbabwe border since the mid-2000s. Source: Pauline Guinard, 2016

Date	*Project*	*Artist(s)*	*Sponsor(s)*	*Type of art*	*Aim of the project*
2006	*Messina/ Musina*	P. Hugo	Unknown	Photographic essay	To question Musina as a place of transition, because of both the end of apartheid and the presence of the border
2007	*Back and Forth*	Students from the Market Photo Workshop	Market Photo Workshop and International Organization for Migration	Photo-reportage	To investigate and raise awareness about the mobility and circulation induced by the border, through retracing the journeys of informal cross-border traders
2008	*Ports of Entry*	R. Watermeyer	Unknown	Photographic essay	To photograph all the places where one can enter South Africa, including the Beitbridge area
2009–2010	*Border Farm*	T. Nkosi	National Arts Council	Docu-drama	To document and understand the migration process as seen and lived by the people living in the border area
2010–2011	*Made in Musina*	T. Nkosi, R. Hlasane	Visual Art Network of South Africa	Festivals and artists network building	To bring together in a network all the artists working in the (South African) border space, and especially in Musina
2013–2014	*Sex Workers Project*	Students from Market Photo Workshop and sex workers from the Musina region	Market Photo Workshop, Centre for Migration and Society, Sisonke Sex Worker Movement	Participative photo-reportage	To explore the sex industry in the border area and to empower sex-workers by training them in photography

or rights, this type of art is also close to what Lacy defines as a 'new genre public art' (1995). This designates a form of art that is not only installed or made in public spaces, but that claims to be made *with* and *for* the so-called public, while initiating public debate. Art in that sense is both relational (Bourriaud 1998) and engaged (Vander Gucht 2004): it aims at having some impact on the social and political context in which it is produced.

At the crossroads of 'spatial turn' in contemporary art and 'new genre public art', border art and artists aspire to act on the spatial, social and political dimensions of the border in order to redefine it as a place of connection rather than of division. But, beyond these claims, one can wonder about the effectiveness of this form of art – especially compared with other economic, social or political dynamics – in (re)uniting the people and spaces at the border. For instance, in a context marked by super-diversity and xenophobia as in the South Africa/Zimbabwe border area, to what extent can a project like *Border Farm* truly and sustainably bring closer people and spaces that tend to be set apart because of the border itself and because of the migration process as a whole?

The *Border Farm* art project: (re)building the self and the group

Border Farm *at the meeting point of an artist's interest and a community's need*

The *Border Farm* project was born from Nkosi's desire to work on the issues of migration, displacement and belonging. As she explains: 'Driven by an abiding personal interest in migrancy and displacement, and how people respond to the pressures of exile, I started researching the situation on the farms in northern Limpopo' (Nkosi 2011: 2). Raised in the United States by a Zulu father and a Greek mother, Nkosi moved to and settled in South Africa in the 1990s, when she was a young adult. Her own personal journey in-between cultures and countries made her sensitive to the migrants' fate. Alarmed by the 2008 xenophobic attacks[2] as well as by the situation of the people living in the South Africa/Zimbabwe border region, she started to gather information on these issues. Then, she decided to organise a first expedition to the border area, assisted by Raymond Marlowe, a photographer. At that time, despite the fact that she was willing to work on the question of belonging, she had no precise idea concerning the type of project she would be conducting there. On the contrary, she 'wanted to see what people wanted' (Interview, January 2014), so that the social and physical spaces of the border would give form and shape to her initial idea.

On one of the farms along the border (Maroi), Nkosi and Marlowe met with M. W.,[3] a Zimbabwean worker. Fortunately enough, Nkosi's idea of working on the notion of belonging almost immediately resonated with M. W.'s own desire to tell his story and those of his fellow farm workers. M. W. had indeed been working on a script about the (migrant) workers' life on the farms along the border. Besides, he shared with Nkosi the feeling that a sense of belonging and community was missing in the area. For both Nkosi and M. W., an art project was seen as a means to bring the farm workers – regardless of nationality, legal status, gender, age, etc. – together, while changing outsiders' main representations on those workers and migrants. The *Border Farm* project was thus born at the intersection of a South African artist's aspirations and a Zimbabwean farm worker's expectations. It was not imposed on the border space or the people living in it, but rather co-produced with and by them. The making of the film itself was conceived as an opportunity to (re)unite the farm workers, while the result of this process (the actual film) was understood as a potential medium able to raise awareness outside the farm, especially amongst South Africans, about the difficulties of the migration process. In order to achieve these goals, Nkosi put in place a collaborative and participative process.

Border Farm: *a participative and therapeutic process redefining the border aesthetic regime*

The project was based on volunteering. During a public meeting organised at the farms in 2009, Nkosi, Marlowe and M. W. presented the project to the entire community and asked the people interested to join in. Twenty-five out of 200 farm workers (mainly migrants but also others) became involved in the project. They formed a theatre group, called Dulibadzimu, which was designed to last longer than the project itself. Nkosi, helped by Marlowe and other artists from Johannesburg, organised workshops in order to help the farm workers develop what was supposed to become *their* project: writing, drama and photography sessions were held in order to elaborate on the script written by M. W. Several months were needed to prepare the shooting that was conducted in January 2010 with and by the farm workers themselves. According to the initiators of the project, the preparation phase was crucial to the cohesion of the group (Nkosi 2011): through the sharing of their personal journeys, the farm workers actually learned from one another. For some, it was even a chance to put into words the traumatic experiences they had endured during their migration. As explained by S. M., a 39-year-old Zimbabwean widow with two children under her responsibility, who illegally crossed the border in 2005:

> I just wanted to be part of that project because I just thought about my life, and the pitiful in it. . . . It was for my remembrance of what I did when I was coming to South Africa. . . . I will never forget. Because when I act the drama, I realised what happened when I was coming. . . . It is a nice project because it teaches us about what is happening in life. The ways… how… some people came, the way they are raped, some of them killed, some of them taken away by the flow.
>
> (Interview, January 2014)

For S. M., the project functioned as a *catharsis*: through the naming and the sharing of what happened to her during the crossing, she could remember and face her own trauma. The preparation and the making of the film were thus a therapeutic process for the farm workers, and especially for the migrants, from which they learned as much about the others as they learned about themselves.

The resulting film[4] recounts, through multiple languages and voices, the stories of the migrants who came to South Africa across the Limpopo River to work on the farms along the border, hoping to find a place where they could sustain their families while staying close to their mother country. The film consists of various sequences, which include and intertwine interviews, footage of the workshops and the re-enactment of the crossing. Interestingly enough, the participants did not necessarily play their own parts. M. W., for instance, chose to be a smuggler instead of a migrant. This reversal of the dominator/dominated roles was for him a way to symbolically regain power over his own life. The film thus contributed to his empowerment. Ultimately, the film as a process was a way for the farm workers to address, individually and collectively, the traumatic experience of migration, both intellectually and physically.

Because the film is about telling the migrants' stories and about showing the re-enactment of their crossing, *Border Farm* offers its participants and its potential viewers an opportunity to better understand but also experience what migration is. It is, as such, a means to explore the extent to which the border affects the migrants' lives and bodies. Indeed, in a post-Westphalian context (Amilhat-Szary and Giraut 2015, Nicol and Minghi 2005), the border can no longer be analysed anymore as a fixed line completely external to the individuals: on the contrary, it has to be understood as mobile and embodied – with various degrees of constraints – by every individual. In that regard, since *Border Farm* gives its viewers a new way of perceiving the reality of the border as it is lived by the migrants, it can potentially change this reality by transforming the representations attached to it. Consequently, it could promote the redefinition of the 'aesthetic regime' of the border, that is to say, according to Rancière (2008), the way the border is seen, felt and experienced.

Nevertheless, despite the therapeutic and empowering function of the film for the participants and its potential consequences on the representations of the border for the participants and non-participants, one can wonder about the impacts of this project in the long term. Was this project successful in its attempt to *sustainably* bring closer the people and spaces divided by the migration process?

Border Farm: *a sustainable reconstruction of the self but not of the group*

In order to grasp the consequences of the *Border Farm* project in the long term, I conducted fieldwork in the South Africa/Zimbabwe border region in July 2013 and January 2014, almost four years after the end of the film making. Because I had never been to that part of South Africa known as a harsh place, I asked Nkosi and Marlowe to come and assist me. Their role was mainly to put me in touch with the people they know in the area and to help me understand the way this area functions. The goal of that fieldwork was to question the extent to which Border

Farm – beyond its relative success[5] – had changed the lives of its participants, the relationships between the migrants and the South Africans in the Maroi farm and beyond, as well as the representations associated with the border.

The material and immaterial participants' personal gains

For many participants, *Border Farm* had – to a lesser or a greater extent – a real impact on their lives, thanks to the skills they learnt or to the therapeutic dimension of the project. Some were indeed able to build on their acting or photography skills acquired during the workshops to start doing new activities. For example, N. N., one of the Zimbabwean farm workers who was particularly interested in photography, had managed to become a professional photographer in a safari resort close to the Maroi farm. Similarly, several participants are still involved in diverse theatre groups – but not necessarily in Dulibadzimu – which are well known by civil society and especially by the NGOs based in the area. For instance, the International Organization for Migration regularly calls on the farm workers' experience in drama to conduct awareness-raising campaigns on HIV/ Aids or domestic violence, two issues particularly widespread in the region. For others, the main gain of the project did not so much rely on the acquisition of new skills but on a certain sense of pride, respect and responsibility the project gave them. Many participants' answers to the question concerning the inputs of the *Border Farm* project on their own lives are symptomatic of that:

> It [*Border Farm*] gave me a sense of responsibility. [...] It gave me self-pride, a sense of purpose.
>
> (E. M.'s Interview, January 2014)

> I feel proud.
>
> (G. N.'s Interview, January 2014)

> I got connected to people. [...] I got respect.
>
> (M. W.'s Interview, January 2014)

These statements confirm the therapeutic function of art for the participants in the long term. *Border Farm* helped the farm workers, and especially the migrants, to create a more positive image of themselves, to increase their self-confidence and self-esteem, while maybe making them more sensitive to others' fate. Surprisingly though, this sense of caring is not necessarily directed towards the former members of *Border Farm*.

The collective challenges for the participants as a group

When Nkosi and the other artists from Johannesburg left the Maroi farm, it became difficult for the group to maintain the collective structures that had been previously established: in 2014, the Dulibadzimu Theatre Group was not active anymore. The link which had been holding the group together during the *Border Farm* project seemed to be broken or at least extremely stretched. The relationships between the members of the group had indeed been complicated by money issues. A sign of its success, the various screenings of *Border Farm* had generated revenues. But the sharing of these revenues caused trouble and conflicts amongst the members of the group. M. W. in particular was accused of keeping some money for himself. This conflicting situation which had been deteriorating the relationships within the group remained latent and unresolved until Nkosi and Marlowe returned.

As mentioned previously, the fieldwork I organised in January 2014 was an opportunity for Nkosi and Marlowe to go back to Musina and to (re)connect with the people they had been working with. Nkosi saw her return as a way to close the *Border Farm* project by bringing the remaining revenues generated by the film to the group. Aware of the rumours concerning the misappropriation of funds, she was quite nervous about the meeting planned with the former participants of *Border Farm*. Nevertheless, the meeting went quite smoothly. As a participant observer, I noticed how the presence of Nkosi and Marlowe fostered the reunification of the group. Knowing they were coming, all the former participants of *Border Farm* who were still working at the Maroi farm met on Sunday, 18 January 2014, at the place they used to do the workshops. While they had not been able to do so for the last two or three years, they all managed to be present for Nkosi and Marlowe. The artists acted as a necessary link in-between the members of the group. And indeed, once they were there, even the money issues that had divided the group seemed to vanish. Through an open discussion moderated by Nkosi, it appeared that the problem came from one participant who had borrowed money from the group and was not able to reimburse. It was then collectively decided that this person would not have a share of the remaining revenues. The issue was thus solved and the reunited group even spoke of making a new film. But how effective could this be without an external presence such as the artists' one?

The capacity of the group to function as one, when the initiators of the *Border Farm* project are not there, compromises its mere existence. The fact that the group decided at the same meeting that Nkosi should stay in charge of the potential future revenues generated by the film – whereas she would have preferred to let someone else from the group and more specifically from the farm do it – is symptomatic of the difficulties of the group to exist without an external (artistic) presence. Is this lack of internal unity due to the super-diversity of the group? Is the border still dividing them despite the attempt of the *Border Farm* project to overcome the discontinuities produced or revealed by it? If *Border Farm* mainly focused on the act of crossing the border that is – directly or indirectly – shared by all the farm workers, this does not mean that the farm workers' experience and life, and the extent to which the border influenced those, are unified. Their nationalities, qualifications, ages or family situations continue to differentiate their abilities to project themselves into the future and into South Africa. In any case, the willingness of its former participants to make a second film is also a sign that for them everything about their condition as migrants and farm workers has not been told yet.

The persisting ambition to change people's representations of the migrants and the border

The interviews I conducted with the members of the group reveal a general change of mindset amongst the farm workers concerning both the meanings given to the film they did and their expectations for a future film.

The meanings the participants gave to their film seem to have been transformed through time. While in the interviews made for *Border Farm* in 2009 and 2010 they expressed a desire to tell their stories to the *South Africans* in order to fight xenophobia and discrimination, in 2014 the farm workers in general and the migrants in particular told me that they actually wished to alert the *Zimbabweans* on the dangers of illegally crossing. If the film was still the same, the message attached to it and its potential recipients had changed in the minds of its participants. How to explain what might appear as reconstruction in the participants' discourses? Furthermore, if the presence of Nkosi and Marlowe activated the *Border Farm* participants' willingness to make another film, what they would like to picture now no longer concerns the act of migration itself but is more about their life on the farm *after* the crossing. Does that mean that migration is not so much at the core of the migrants' preoccupation? What does that tell us about their current situation?

These changes of mindset reflect a transformation of the migrants' situation in the border region, at least at the time I did the interviews. In 2014, thanks to a progressive improvement of the political, economic and social situation in Zimbabwe, fewer people were illegally crossing the border. Xenophobia had also tended to be more latent than active in the border area and in South Africa. The migrants, and especially those who chose to stay in the border area, were slowly settling in. The main difficulties they faced were related to their daily life within the border region: low wages, poor housing, domestic violence, HIV/AIDS, etc. The changes of concerns expressed by the *Border Farm* former participants about the potential role of art in the region have thus to be understood as a consequence of the evolving socio-spatial context these people are living in.

As a result, the function and the nature of art would have to adapt to this changing context in order to respond and resonate again with it. If *Border Farm* had corresponded to what the border was at the time the film was made, if it could have helped its audience to better grasp its complexity during that period, it was not necessarily the case anymore. As the border is dynamic and mobile (Amilhat-Szary and Giraut 2015), so should be border art. In that respect, art might not only be a means to address a particular socio-spatial border context but it might also be the diagnosis of a situation (Smith 2015), a tool to reveal what the border is and how it evolves through time.

Art as object and tool to apprehend the spatiality of the border

To study the border in its materiality and immateriality, as it is and as it is lived, art can be both an object and a tool for geographers. On one hand, the geography of the art projects that took place in the South Africa/Zimbabwe border area can be a means to apprehend the border not (only) as a fixed line materialised in space by fences and barbed wires but (also) as a mobile, multiple and dynamic space. On the other hand, art can be used by geographers as a way of doing (Hawkins 2011), that is to say as a tool and a practice in order to comprehend the border as it is perceived and experienced. As underlined by Lefebvre (2000), art is indeed crucial for understanding the production of space, because of its capacity to act on the concrete and symbolic dimension of space.

A mobile and multiple border space

The cartography of the border art projects made in the South Africa/Zimbabwe border area (Figure 10.1 and Table 10.1) and their analysis, bring to light the fact that the spatiality of that border is actually more complex than it might appear at first. Since the presence of the river, fences or barbed wire makes it highly visible, this border seems to be strongly rooted in space and easy to locate. The photographic essay, *Ports of Entry* by Robert Watermeyer,[6] conveys this idea by offering very fixed images of the border in its materiality. But, by looking at other arts project like *Messina/Musina*, *Border Farm* or *Back and Forth*, the border appears much more unstable and moving. The presence of the South African soldiers, notably photographed by Pieter Hugo in *Messina/Musina* (2007), reveals the mobile nature of the border which is extended through the soldiers' patrols in the area. Similarly, the focus on the migrants in *Border Farm* or on the cross-border traders in *Back and Forth* gives visibility to the extent to which the border is integrated into the migrants and traders' bodies and how it consequently circulates with them. Through art, it is thus possible to unveil the mobile nature of the border (Amilhat-Szary 2015).

Furthermore, the various art projects that were produced in and with this border space and especially the two projects initiated by Nkosi, *Border Farm* and *Made in Musina*, make the multiplicity of the border apparent. Through her own (artistic) practice of the area, Nkosi realised that there was another border – surely less visible but effective – that was structuring the region: the one in-between the farms established along the actual international border and the urban spaces composing the small town of Musina. This border is as much economic, social and spatial, as it is symbolic. The two spaces along it correspond to two forms of socio-spatial organisations that tend to be respectively rural and urban. Their demographics differ by the percentage of migrants amongst their inhabitants (majority of Zimbabweans on the farms and of South Africans in town). Moreover, the lack of affordable transportation to go from the farms to town, reinforces the real and symbolic distance that exists between these spaces. But, unlike the actual South Africa/Zimbabwe border that draws the line between different nations, this social, economic, spatial and symbolic border is not strictly speaking based on the nationality of the people living in these spaces since there are – even in small proportion – South Africans on the farms and foreigners in town. This division is rather defined by the proximity to the international border, by the extent to which this border affects the people's lives and how they engage with it. Eventually, what seems to separate these two spaces is their different relation to the international border, their 'borderity', as Amilhat-Szary and Giraut would say, that designates 'both an individualized and a collective relationship' that participate 'in the making of a differentiated and individualized border' (2015: 11). As such, the border separating the farms from the town is not really another border, completely different from the South Africa/Zimbabwe one. It is more accurately an extension and a duplication of the latter. It materialises the expansion of the South Africa/Zimbabwe border within the South African territory, and by doing so, it reveals the thickness of the border and its zonal (rather than linear) nature. Yet, the South Africa/Zimbabwe border considered in its depth is not a uniform zone. For instance, the mechanisms of controls associated with the border change depending on the distance from the official borderline: while moving away from the border post, mobile and random controls replace fixed and systematic ones.

It is precisely because *Border Farm* made these multiple divisions apparent for Nkosi that she decided to do another project in the region: *Made in Musina*. This project started from her will to bring the farms and the town closer together. But despite her initial goal, *Made in Musina* failed in challenging this socio-spatial separation and finally became a Musina project almost exclusively (Guinard forthcoming). Contrary to the artist's intentions, by associating – even through their name – one project to one space, *Border Farm* and *Made in Musina* had tended to confirm the existence of the socio-spatial division between the farms and the town. In that case, art did not only make the border visible, it also reinforced it. As stressed by Rancière (2008), presupposing the effects of art is illusory and vain. This does not mean that art is ineffective or pointless, but it emphasises the fact that the effectiveness of art lies in its openness, in its uncertain ability to modify the ways one feels and experiences the world. Nevertheless, in spite of – or thanks to – this uncertainty concerning its effects, art can still appear as a relevant means to understand the reality as it is and as it is perceived.

The border as lived or abstract space

Based on Lynch's methodology of mental maps (1960), I used art, and in that case drawing, as a methodology in order to bring into light the ways the people living in the South Africa/Zimbabwe border area, and especially those who had been involved in art projects, perceived the border. I asked the 15 persons I interviewed on the farms, the 14 persons I interviewed in town and the 9 children and teenagers I did a focus group with[7] to 'draw the border'.

Regardless of the nationalities, genders and ages of the interviewees, three main elements structure the 38 mental maps produced: the Limpopo River; the Beitbridge Bridge; and the road linking South Africa to Zimbabwe. The river is prevailing in the farm workers' representations, even if they are not border jumpers themselves, like H. S. (Figure 10.2a). Because the farm workers live close to

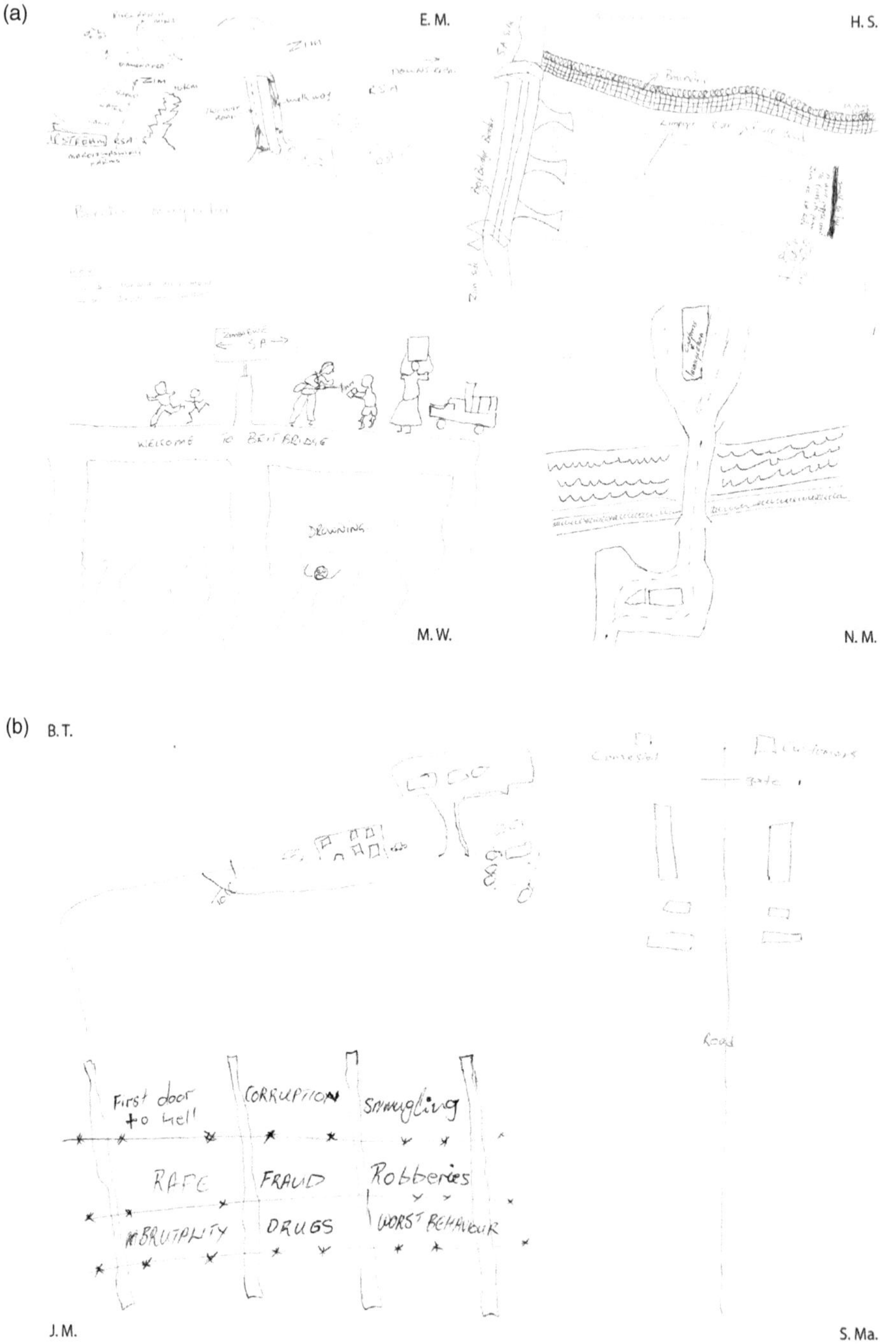

Figure 10.2 (a) The border as represented by the farm workers, a lived space; (b) The border as represented by the people living in Musina, an abstract space. Source: Pauline Guinard, 2016.

the river and regularly witness illegal crossings, they perceive and represent the border as dual: the 'official border' on one side and the 'river border' on the other side as they call them. In their drawings, they give visibility to the border as it is represented and practised.

On the contrary, even if the same three elements (the river, the bridge, the road) structure the maps drawn by the people living in Musina, they almost exclusively represent the 'official border' (Figure 10.2b). This is particularly true for South Africans.

Furthermore, the representations of the border made by the inhabitants of Musina are often more abstract than the ones made by the farm workers. Because the people, and especially the South Africans, living in Musina do not often go across, the border and its surroundings remain an unfamiliar territory, a blurred space. Many of them were even reluctant to produce a drawing, arguing that they did not really know the border. In most of their mental maps, the closer to the actual border it gets, the less detailed the drawings are, while the Zimbabwean side of the border is frequently missing.

Conversely, the people living on the farms as well as the migrants living in Musina know the border better because they have – directly or indirectly – experienced it. In their case, drawing the border often prompted memories about a crossing by themselves or others. As a result, they have a more personal way of representing it. The mental map drawn by M. W. is particularly striking. He depicted the two borders, the river one and the formal one, but also their consequences for the people using them: while some people are rather quietly crossing through the bridge legally, one man who is trying to cross through the border is drowning. This drawing vividly illustrates the inequalities the people could experience – including through their bodies – at border sites. It gives form and life to these experiences. As a consequence, mental maps appear as an enriching tool to apprehend in an alternative way how the border is perceived and lived by someone who had experienced it at a greater or lesser distance. More generally, mapping and especially participative mapping involving artists and migrants is currently explored by some geographers (Mekdjian et al. 2014), so as to find new ways of telling migration as it is lived by migrants themselves. As an often non-verbal, collaborative and participative tool, art could thus be a new medium for geographers, to examine the relationships the people – individually or collectively – entertain with space in general and with border space in particular.

Conclusion

As stressed above, the South Africa/Zimbabwe border area is – as in the case of many borders around the world (Nicol and Minghi 2005) – a contradictory space. Its contradictory nature is surely part of what attracts artists, who are not necessarily from this area, to this region to conduct projects. The complexity of the border stimulates their imagination. As such, the border acts as a catalyst for art production at least on a regional scale and especially in times of crisis. When the border is secured and reinforced, artists tend to be drawn towards it in an attempt to question or challenge this process of re-bordering (Amilhat-Szary and Giraut 2015) and its consequences on the people living there.

Border Farm was a project of that kind. It was conceived by Nkosi as a means to contest the dividing nature of the border by bringing people together through the sharing of the migrants' stories. The border was conceived as a potential place of connection between people, irrespective of their nationality, instead of a site of division. And indeed, *Border Farm* succeeded in addressing some personal trauma lived by the border jumpers as well as in (re)uniting the participating farm workers, be they migrants or not, during the time of the project. This highlights the potential therapeutic function of art for both the individuals and the group, even when the group is notable for its super-diversity. Nevertheless, the impacts of this project are more uncertain in the long term and on a broader scale. This underlines the need for researchers to consider such projects on different spatial and temporal scales. In the *Border Farm* case, the collective momentum had been declining with the departure of the artists who initiated the project. The presence of the artists thus appears as an external stimulus necessary to the cohesion of the group. This raises the question of the possibility for a project like this, which is initiated by an outsider, to generate a sustainable endogenous collective process. Furthermore, the capacity of that kind of art project to have an impact beyond the group itself is more than uncertain. If a study of the reception of *Border Farm* in South Africa and beyond is still to be done, what is already striking is that the dialogue between the people living on the farms and the ones in Musina does not seem to have been facilitated by the film. This might be because the people in Musina have not seen it, but this might also be because the film was not able by itself to overcome the distance separating the two groups of people. On the contrary, by giving visibility to the separation in-between the farms and the town, *Border Farm* and *Made in Musina* (Nkosi's following project) might have made this division more tangible.

However, if the capacity of art to change the (border) context must not be presupposed, that does not mean that art is void of effect (Rancière 2008). Because of its symbolic dimension and of its potential to act on people's perceptions and affects (Thrift 2004), art can be a means – for artists and geographers – to explore the reality as it is and as it is perceived, represented and lived (Lefebvre 2000). Actually, *Border Farm*, as with other art projects produced in the South Africa/Zimbabwe border area, made the complex, mobile and multiple nature of the border visible. Far from the representations that apprehend the border as a fixed and stable line, most of these projects investigated the border as it is lived. This contributes to giving the border a very different image: one of a space that is contested, unstable and moving. By doing so, these art projects – individually and collectively – have also been participating in changing the 'aesthetic regime' of the border (Rancière 2008). Studying these projects could thus be a way for geographers to have a better understanding of how (border) space is produced, including through art. But as mentioned previously, art is not only an object of study for geographers; it might also be a methodological tool. Using mental maps in such a context was actually a way to grasp the border as it was experienced and represented by the people who are living within this socio-spatial context and are affected by it. Ultimately, be it as an object or a tool, art appears as a relevant means to analyse the production of (border) space in its materiality and immateriality.

Acknowledgements

This research was undertaken within Amilhat-Szary's research project 'Mobile Borders' (an Institut Universitaire de France program).

Notes

1 The first (1880–1881) and second (1899–1902) Boer or Anglo-Boer wars brought the British and the Boers (descendants of Dutch settlers) into conflict over the control of what is now the Republic of South Africa.
2 The 'xenophobic attacks' refer to a wave of lethal violence against predominantly – though not exclusively – Black foreigners in South Africa (Wa Kabwe-Segatti 2008). It started in May 2008 in the Johannesburg region before spreading across the country.
3 To respect the general code of conduct regarding the identity of research subjects, especially of marginalised populations such as (illegal) migrants, I anonymised the participants' names.
4 https://vimeo.com/72422149.
5 The film was quite widely released. Nkosi presented it as part of festivals and conferences in South Africa, Zimbabwe and Europe. It was even broadcast on Soweto TV, a popular South African community television.
6 www.robert-watermeyer.com/ports-of-entry/.
7 All the parents of the underage participants signed a consent form for the focus group.

References

Amilhat-Szary, A.-L. 2012. Walls and border art: the politics of art display. *Journal of Borderlands Studies*, 27(2), 213–228.

Amilhat-Szary, A.-L. 2015. *Qu'est-ce qu'une frontière aujourd'hui ? [What is a border today?]*. Paris: Presses Universitaires de France.

Amilhat-Szary, A.-L. and Giraut, F. (eds) 2015. *Borderities and the Politics of Contemporary Mobile Borders*. New York: Palgrave Macmillan.

Andreas, P. 2003. Redrawing the line: borders and security in the twenty-first century. *International Security*, 28(2), 78–111.

Berelowitz, J.-A. 1997. Conflict over 'border art'. Whose subject, whose border, whose show? *Third Text*, 11(40), 69–83.

Bourriaud, N. 1998. *Esthétique relationnelle [Realational aesthetics]*. Dijon: les Presses du réel.

Cosgrove, D. 2008. Cultural cartography: maps and mapping in cultural geography. *Annales de géographie*, 660–661(2), 159–178.

Didiot, M. 2013. Les barrières frontalières : archaïsmes inadaptés ou renforts du pouvoir étatique? [Border barriers: inappropriate archaisms or State power reinforcement.] *L'Espace Politique*, 20. Available at: <http://espacepolitique.revues.org/2626> [accessed: 6 August 2013].

Guinard, P. Forthcoming. Redefining publics, artists and urban spaces: the case of *Made in Musina*. *City & Society*.

Hall, T. 2003. Art and urban change: public art in urban regeneration, in *Cultural Geography in Practice*, edited by A. Blunt. Oxford: Oxford University Press, 221–237.

Hawkins, H. 2011. Dialogues and doings: sketching the relationships between geography and art. *Geography Compass*, 5(7), 464–478.

Hugo, P. 2007. *Messina Musina*. Rome: Punctum.

Janczak, J. 2011. *De-Bordering, Re-Bordering and Symbols on the European Boundaries*. Berlin: Logos Verlag Berlin GmbH.

Lacy, S. (ed) 1995. *Mapping the Terrain: New Genre Public Art*. Seattle, WA: Bay Press.

Lefebvre, H. 2000. *La production de l'espace [The production of space]* (4th edn). Paris: Anthropos.

Lehec, C. 2014. *Fabriquer du paysage. Une expérimentation artistique et géographique autour du mur de séparation à Jérusalem et Bethléem [Making landscape. An artistic and geographic experimentation along the Jerusam/Betheem separation wall]*. Master 2. Grenoble: Université Joseph Fournier.

Lynch, K. 1960. *The Image of the City*, Cambridge, MA: MIT Press.

Mekdjian, S. et al. 2014. Figurer les entre-deux migratoires: Pratiques cartographiques expérimentales entre chercheurs, artistes et voyageurs [Representing the migratory in-betweens: experimental cartographic practices between researchers, artists and travelers]. *Carnets de géographes*, 7. Available at: www.carnetsdegeographes.org/carnets_terrain/terrain_07_01_Mekdjian.php [accessed 31 July 2015].

Newman, D. and Paasi, A. 1998. Fences and neighbours in the postmodern world: boundary narratives in political geography. *Progress in Human Geography*, 22(2), 186–207.

Nicol, H. and Minghi, J. 2005. The continuing relevance of borders in contemporary contexts. *Geopolitics*, 10(4), 680–687.

Nkosi, T. 2011. Return to Border Farm. In 'wide angle: photography as public practice' forum. Johannesburg.

Rancière, J. 2008. *Le spectateur émancipé [The emancipated spectator]*. Paris: La Fabrique.

Rosière, S. and Jones, R. 2012. Teichopolitics: re-considering globalisation through the role of walls and fences. *Geopolitics*, 17(1), 217–234.

Sheren, I. 2015. *Portable Borders: Performance Art and Politics on the U.S. Frontera since 1984*. Austin, TX: University of Texas Press.

Smith, C. 2015. Art as a diagnostic: assessing social and political transformation through public art in Cairo, Egypt. *Social & Cultural Geography*, 16(1), 22–42.

Stevens, Q. 2007. *The Ludic City: Exploring the Potential of Public Spaces*. London, New York: Routledge.

Thrift, N. 2004. Intensities of feeling: towards a spatial politics of affect. *Geografiska Annaler: Series B, Human Geography*, 86(1), 57–78.

Vander Gucht, D. 2004. *Art et politique: pour une redéfinition de l'art engagé [Art and politics: for a redefinition of engaged art]*. Bruxelles: Labor.

Vertovec, S. 2007. Super-diversity and its implications. *Ethnic and Racial Studies*, 30(6), 1024–1054.

Volvey, A. 2007. Land Arts. Les fabriques spatiales de l'art contemporain [Land Arts. The spatial making of contemporary art]. *Travaux de l'Institut de Géographie de Reims*, 129–130, 5–27.
Volvey, A. 2014. Entre l'art et la géographie, une question (d')esthétique [Between art and geography, an issue of aesthetics]. *Belgeo*, 3. Available at: http://belgeo.revues.org/13258 [accessed 26 May 2015].
Wa Kabwe-Segatti, A. 2008. Violences xénophobes en Afrique du Sud: retour sur un désastre annoncé [Xenophobic violence in South Africa: chronicle of a disaster foretold]. *Politique africaine*, 112, 99–118.
Wilson, T. and Donnan, H. 2012. *A Companion to Border Studies*. New York: Wiley.
Zebracki, M. 2011. Beyond public artopia: public art as perceived by its publics. *GeoJournal*, 41, 786–795.

11 The Birmingham Surrealist Laboratory

Unlocking community and the avant-garde in a super-diverse city

Saskia Warren and Stephen Forcer

Introduction

> A cloud counting hot butchers.
>
> (Exquisite Corpse: I; Workshop 2014)

> Policeman chasing angry door.
>
> (Exquisite Corpse: II; Workshop 2014)

> Do not judge this movement kindly. It is not just another amusing stunt. It is defiant, the desperate act of men too profoundly convinced of the rottenness of our civilisation to want to save a shred of its respectability.
>
> (Herbert Read, International Surrealist Exhibition Catalogue, London 1936, p. 13)

> The English are still the maddest people in the world. Already the whole of Europe considered they were crazy because they live on a humid island, through choice it seems, they eat copiously as soon as they get out of bed, they sunbathe whenever they can, they kill foxes in great ceremony, and cry their eyes out at the death of a dog.
>
> (Henry Treece 1941)

> If it's private and inside then I don't think anyone would be offended.
>
> (Participant, Workshop II 2014)

The Birmingham Surrealist Laboratory investigates the ways in which new digital facilities can help unlock complex issues of art, community and cultural sensitivity in a super-diverse city. Its starting point is the little-known history of the Birmingham Surrealist Movement of 1935–1955 that produced mixed media work including critically recognised collage, paintings, poetry and performance work. Despite an important heritage within one of the most alluring and recognisable artistic movements of the twentieth century, comparatively little academic and wider public attention has been afforded to the Birmingham Surrealist Movement (Sidey 2000). In this chapter we revisit the overlooked heritage of the Birmingham Surrealist Movement in the context of contemporary Birmingham. It outlines an experimental project whereby the history and work of Birmingham Surrealism is introduced to communities who today live and work in Balsall Heath; the neighbourhood in which the original members collected, debated, played and produced avant-garde art.

Using multi-touch digital facilities at the cutting-edge Digital Humanities Hub, University of Birmingham, the project tested the ways in which Surrealism travelled to community participants enrolled from Balsall Heath. The project comprised two workshops that introduced Birmingham Surrealism, including Surrealist word games, interactive digital play with Surrealist images and a round-table discussion on the legacy of Birmingham Surrealism. It aimed to explore the ways in which the symbolism of Surrealist art – replete with violent, sexual and often irreligious imagery – is encountered by those from different cultural and religious backgrounds, and to use digital technology to open up debate on community heritage. Together these aims interrogate issues of access to widening engagement with the heritage of the Birmingham Surrealism movement, teasing out tensions of physical and perceptual barriers to access that are unevenly and differentially experienced across social variables of culture, religion, socio-economic background, and educational attainment.

As a neighbourhood Balsall Heath can usefully be framed as super-diverse – with a population in flux characterised by in-and-out migration from newer, more stratified individuals and groups; and distinguished from the flows of previous immigrant arrivals by more wide-ranging migration routes, legal statuses, intended length of stay and intersectional markers of ethnicity, culture, religion and socio-economic background (Meissner and Vertovec 2015, Warren and Jones 2015). It is within this context of fast-changing social, material and religious relations that we seek to locate experiences, resonances and values of British avant-garde art and its legacies.

The Birmingham Surrealist Movement (1935–1955)

This research investigates communities of interest in the cultural heritage of Birmingham Surrealism, part of the wider European Surrealism movement. Surrealism can be considered a philosophy of life and art. Those involved in its inception sought to reject rationalism, perceived as a kind of European societal malaise that had led, step by step, to the violent cataclysm of World War I. The Surrealist Manifesto was published in 1924, with André Breton, its intellectual leader, defining the movement as 'Pure psychic automatism, by which it is intended to express, verbally, in writing, or by other means, the real functioning of thought' (1978: 122). As Bonnett (1992: 74) has written, the surrealists placed importance on the 'exploration and liberation of the unconscious through the development of forms of cultural production that privileged spontaneity and chance'. The spontaneity of automatism can be observed across media from surrealist poetry, collage, painting, film and, later, the surrealist-inspired Parisienne *dérives* of the Situationist Internationals in the late 1950s and 1960s (Adamowicz 1998, Bonnett 1992). Paris was at the centre of the European Surrealist movement; however, Surrealist networks were manifest globally with distinctive movements in parts of Scandinavia, South America, Egypt, Japan, Mexico and the Carribean (Mostafa 2005: 216).

English surrealism was announced in 1935 with the publication of the 'First English Surrealist Manifesto' by poet David Gascoyne in the French journal *Cahiers d'art*, with the founding of the Birmingham Surrealist Group following in 1936. Relatively little is written on the Birmingham group and its leader Conroy Maddox. We know that Maddox, originally from Ledbury, discovered European avant-garde and Surrealism in particular during his early twenties at Birmingham City Library and through *Cahiers d'art* and *Le Surréalisme au service de la révolution* at Zwemmers Gallery, London (Adamowicz 2014). Maddox later rejected the London Surrealist scene, refusing to exhibit, along with brothers John and Robert Melville, in the high-profile 1936 International Surrealist London exhibition due to a belief that the 27 artists selected were diluting the principles of Surrealism. Maddox considered several of the artists exhibiting were identifying themselves as Surrealist for the purposes of forwarding their own careers. Typically the Surrealists had ascribed to collectivism with committed artists showing in groups rather than individually. For example, Merlyn Evan's *Conquest of Time* (1934) is an oil painting closer in style to Vorticism and Cubism. Meanwhile, Sir Roland Penrose's *The Last Voyage of Captain Cook* (1936) is formed of plaster and steel on a wooden base and more accurately an assemblage. But, despite his refusal to exhibit, Maddox attended the seminal exhibition launch, and there he met André Breton and Salvador Dalí with whom he maintained close contact through regular letter-writing and personal visits.

Maddox and the Birmingham Surrealist Group positioned themselves as 'hard' Surrealists who 'were always on the side of Breton' in revolutionary spirit (Adamowicz 2014). In particular, Maddox was drawn to the 'internal revolution', a revolution of the mind that externalised what was inside, placing it on paper. As such his artistic approach sought to bring private dreams and passions into public view. These were dangerous aspirations within a well-to-do, mainstream, socially conservative English cultural milieu as lamented by Maddox and John Melville in the letters they exchanged in the *Birmingham Post* in 1935 (Forcer 2013). A 'Surrealist King Pin' (Forcer 2013), Maddox convened an assorted group of communist intellectuals from the University of Birmingham, along with artists and musicians at a now-demolished house at 29 Speedwell Road, Balsall Heath (Remy 2001, Sidey 2000). The Birmingham Group comprised Maddox, Desmond Morris, John and Robert Melville, Oscar Mellor, William Gear, Eric Malthouse and Emmy Bridgwater. It was at the centre of a community of alternative cultural figures including jazz musician George Melly, writers Stuart Gilbert and Henry Green, and poet Henry Reed. Perhaps most regarded for his collage and paintings, Maddox's work is held by Tate Collection – *Untitled drawing ('In Paris')* (1937); *The Strange Country* (1940); *Winter Criminal Term* (1963); and *Passage de l'Opéra* (1970–1) – indicating his contribution to the British art canon. Maddox's fervour for orthodox surrealism aligned the Birmingham group with the core Parisian Surrealists rather than their London-based counterparts. It has been argued, therefore, that Birmingham was the authentic locale of British Surrealism between the 1940s to 1950s (Levy 2003).

Balsall Heath: an unlikely hotbed of Surrealist activity?

We next explore the ways in which the curious heritage of the Birmingham Surrealist Movement translates to contemporary residents and workers in Balsall Heath. In the experimental Birmingham Surrealist Laboratory project we used new digital facilities that could harness and share materials otherwise dispersed in different collections, both public and private. In two public-facing workshops we investigated how the avant-garde is encountered by our participants. Two main events acted as the points of departure for the workshops. First, a Surrealist House competition staged as part of an art programme, *Balsall Heath Biennale*, intended for those resident in the area of Balsall Heath (2012–13). *Balsall Heath Biennale* was organised by two resident artists who reworked the idea of a conventional biennale – an event *occurring* every two years – by instead stating an arts-based commitment to the neighbourhood for a *duration* of two years. Second, the National Trust at the time was interested in funding a new heritage space in Birmingham. These events acted as a catalyst for the project where we aimed to support better understanding of the legacy of Birmingham Surrealism for the multiple communities of Balsall Heath, including the majority British-Pakistani Muslim population, and to consider its potential future manifestations.

Representations of Balsall Heath tend to be far removed from rumination on the symbolic gestures of the European avant-garde. It is well documented that Balsall Heath has been identified as falling within the lowest 5 per cent of neighbourhoods in the UK for multiple deprivation (Census 2011). The area has been named as one of 25 Birmingham City Council's Priority Neighbourhoods – a formal area below the level of wards comprising around 15,000 people – meaning it is identified as especially disadvantaged. Located 2.5 miles south of the city centre within Sparkbrook ward, residents of Balsall Heath recorded their religion as 70.2 per cent Muslim, and ethnicity as 60 per cent of Asian origin, 12 per cent white, 11 per cent black, and 10 per cent as 'Arabs' (Census 2011). Predominantly Muslim and British Pakistani, these statistics nevertheless mask an ever-changing population with recent arrivals including Somalian, Iraqi, Iranian, Afghani and Chinese, alongside more established Bangladeshi and Yemini communities. Challenges of worklessness are acute with 36.5 per cent economically active and in employment, and no adults in employment in 41 per cent of households (Warren and Jones 2015: 1744). However, Balsall Heath must be contextualised in relation to its more affluent neighbouring areas of Edgbaston and Moseley. The neighbourhood has attracted a number of leading art organisations and artists that have organised and delivered programmes there, attracted by its vibrancy, diversity and, also, its relative proximity both to the city core and residential areas where high numbers of professionals reside. These programmes have tended to be characterised by the aim of bringing together different ethnic and religious groups (for example, Balsall Heath Biennale, Balsall Heath Carnival, Ort Gallery and Near Neighbours). It is further noteworthy that the neighbourhood acted as stimuli for the Conservative political party notion of 'The Big Society' where it was represented as a model for community-led action, with active forums, trusts, a local library and annual carnival.

Fostering inter-cultural exchange has proved challenging, however. Cultural engagement from the non-white population and young people in Birmingham is significantly below the UK average (–11 per cent) (Big City Culture 2010–15, Birmingham City Council). The low take-up from the non-White resident population in the Balsall Heath Biennale Surrealist House competition gave pause for reflection on which variables were impacting upon disengagement with mainstream publically funded cultural programming. Digital has often been heralded as a medium that can activate change at the scale of the individual and society (Elwood and Leszczynski 2013). Despite claims for the transformative potentialities of digital, sometimes verging on the utopian and apolitical, empirical studies have not investigated these claims in relation to so-called hard-to-reach communities.

Digital technologies can be considered to offer a new tool through which to investigate multiple barriers to mainstream, and more subversive, manifestations of culture and heritage in the city. Bringing together a spatial encounter between digital interfaces, subversive artwork and participants from diverse backgrounds would potentially unsettle the ground of established ways of doing. This experimentation adds further weight to progressive conceptions of a sense of place (see Massey 1994) that challenges 'the misleading, tokenist and reifying view of communities as never-changing, socially bounded entities' (Vertovec 2010: 85) that critics have argued is imbued in multicultural ideology.

'In recognition that communities themselves have changed', Raco and Flint argue for the need to acknowledge 'greater emphasis on communities of identity and interest which are not necessarily centred on places' (2001: 588). Bringing together communities of identity and interest with discourse and patterns of super-diversity renders it particularly clear that a community means more than sharing a neighbourhood. We might think of a community of interest that can either be defined in racial and ethnic terms that 'transcends space as a result of the group's shared history, culture, sense of group identity, legacy of discrimination and segregation' (Leib 1998: 688), *or* as a 'formal and/or functional region which can cut across racial and ethnic group divisions' (1998: 683). It is this latter definition of community of interest as a functional region that can intersect social division that is taken forward as an analytic framework for the present research. Shared geographical interests for diverse residents and their common curiosity in exploring the Surrealist heritage of their neighbourhood appealed across social markers of racial, ethnic and religious difference, and educational and socio-economic background.

Research questions

Informing innovative issues of cultural education and the avant-garde in a diverse city, the project data feeds into wider research on how participants map meaning from digital interfaces in heritage and museum spaces. The project built upon existing community relationships developed as part of a larger AHRC Connected Communities project, *Cultural Intermediation*, and *Balsall Heath Biennale*. It culminated in a roundtable format discussion drawing community participants, cultural sector workers and multi-disciplinary academics from Art History, French Studies, Film Studies and Cultural Geography to ask new research questions on complex issues of cultural transmission and cultural sensitivity in the legacy of Surrealism in Birmingham, and its wider global movements (Barber Institute 2014).

The wider project was used to hold debate, test a concept and develop partnerships, with three core imperatives:

1 Improve understanding about how the often irreligious and overtly sexual and violent symbolism of Surrealist art translates to diverse community members from different ethnic, cultural and religious traditions.
2 Investigate the cultural barriers to low (mainstream) cultural engagement from the non-white population and young people in Birmingham (Big City Culture 2010–15, Birmingham City Council).
3 Open a dialogue with participants on what the living heritage of Birmingham Surrealism means in Balsall Heath today.

Surrealism and the Digital Humanities Hub

Multi-touch digital facilities were used at the Digital Humanities Hub, University of Birmingham, to test the ways in which Surrealism translates to multi-cultural community groups recruited from Balsall Heath. Uploaded onto the screens were an assortment of images that depicted formal Surrealist artworks, everyday Surrealism (for instance, a well-known house in Oxford with a model Shark crashing through its roof); the artistic and social life of Conroy Maddox; and the wider Birmingham Surrealist Movement and urban icons from Balsall Heath, such as street signage and the local park. Using Pixelsense touch recognition, participants could then play with these images: either zooming in to see any detail with greater precision; visually or thematically grouping images according to interest; or making them into a collage. Participants were able to make use of 4 × vertically mounted 65" screens, and a 3m × 2m 3D multi-touch wall. Interactive elements were accompanied by information on surrealism and how the concept has been explored in myriad artistic mediums and cultural contexts.

The multi-touch tables are designed to work with heritage organisations to engage new audiences and improve the learning experience. Subject matter can be explored through an array of images that appear randomly scattered. Each image can be 'dragged, rotated, zoomed and shrunk by multiple users simultaneously' (DigitalHumanitiesHub 2015). The hub represents part of the paradigmatic shift towards the new *digital* humanities that attends to transformation in cultural production, distribution and audiencing taking place with and through digital technologies (see Rose 2015: 6). Particularly effective for facilitating multi-user rather than single-user experience, the interfaces can convene a number of different users in one space requiring interaction and negotiation between participants (HIVE, Worcester, Museums and Heritage Awards 2013). An 'ecology' (Ash 2015, Rose 2015) is generated of hardware (the screen, its physical components), software code (image file and operating system) and, crucially, the agency of users.

The interface creates a virtual, interactive collage. Elza Adamowicz has described the Surrealist collage aesthetic as 'radical forms of representation based on the transformation of pre-existing elements' (1998: 3). Surrealist collage is conceived as disrupting the social order by a practice that 'deliberately subverts traditional models of representation and bourgeois value systems, through strategies of displacement and perversion' (1998: 11). By juxtaposing disparate images, then, the technique destabilises their usual meaning and can offer a parodic reworking of narrative (and social) clichés.

Despite liberatory gestures within Surrealist art theory and practice, and digital humanities, working with the avant-garde and digital together can nevertheless be held in tension with widening access to culture. While the Surrealist movement, in its various stages and manifestations in different locales, celebrates the universalities of desire, dreams and the imagination, the material form (painting, sculpture, film, poetry, performance) utilise a highly specialised, if not elitist, conceptual symbolism. Breton, 'a poet and theorist above all', was seeking to establish the foundations of an aesthetic philosophy alongside raising Surrealism's standing as an artistic movement (Polizzotti in Breton 2002: xvii, xix). Digital humanities have typically positioned digital with facilitating more democratic learning processes through enhancing access, distribution, and ultimately participation. But despite hands-on, interactive potentialities, perceptual barriers remain for accessing new digital technologies, especially across axes of difference. Attending a workshop in a digital suite at a university on Surrealist heritage undoubtedly creates a number of perceptual, if not material, barriers to facilitating common debate. It must therefore be recognised that a select community of interest was revealed in the study; those who responded to adverts and were able to convene in the Hub.

Workshops

For the workshops participants were recruited using a snowballing technique from existing community contacts developed across the *Cultural Intermediation* project. Recruitment techniques involved desk-based research, site visits and participation in local forums, supplemented with using local media platforms including an advert for participation placed in the free local newspaper, *Balsall Heathen*. In line with the approach of the project – that participants are experts in relation to the area in which they live and work, and own internal consciousness – time given to involvement in the workshops was remunerated. There were high levels of interest, with both workshops heavily oversubscribed. The first 20 participants who booked onto each of the workshops were offered a place (40 in total). While avoiding intrusive questions into social background – bearing in mind that participants may already feel out of place within the university setting – the demographic make-up of the participants was highly diverse from signs given by their names, and appearance, including ways of dressing, speaking, age and race. In the workshops, participants were introduced to the digital technology of the Digital Humanities Hub, Surrealist mind games, the Birmingham Surrealist Group and the global movements of Surrealism. The 'laboratory' was participant-led using the multi-touch tables in the Digital Humanities Hub. However, the researchers were present to explain the technology, guide the workshops and record the data. Each laboratory session closed with a reflective discussion on Surrealism in the context of Balsall Heath and its communities, past and present.

Exquisite Corpse: *Surrealist mind-games*

To explore the workings of the mind, the new and unanticipated, Surrealists experimented with word games, hypnosis, drug-taking, sleep deprivation and speaking in tongues to support the production of various artistic mediums, including poetry, painting, film, collage and music. In the workshop participants were introduced to 'the expression of the fragmented inner voice' (Adamowicz 1998: 6), through the game of *Exquisite Corpse* (Figure 11.1). The rules of the game were explained by one of the researchers: player one names an object, player two names a verb, player three names an adjective and player four names a noun. The resulting sentence creates an often strange juxtaposition of imagery that places the familiar into unfamiliar scenarios. These results were compounded by participants freely interpreting what constitutes an object, noun, verb and adjective. *Exquisite Corpse* produced the following wordplay in the workshops:

- Yesterday's supper is drinking textured a lightbulb.
- Goldfish destroy pampered rabbit.
- Googlemaps gambled sharp sneaky.
- Chimera create hesitant tile.
- My Persian cats are drying planet.
- Corpse running beautiful elbow.
- Policeman chasing angry door.
- A cloud counting hot butchers.
- Painting opens cold rainbow.
- Cabbage earth travelling crucifixion.

Exquisite Corpse acted as an effective ice-breaker with laughter from the participants and workshop coordinators arising out of the transformation of the banal into the fantastical as the sentences unfolded in unexpected, and often non-sensible ways. More relaxed, the participants were next asked to work in pairs or small groups around the mounted interactive screens (Figure 11.2). Apart from images relating to the everyday material culture of Balsall Heath and some recognition of the paintings of René Magritte, most of the participants

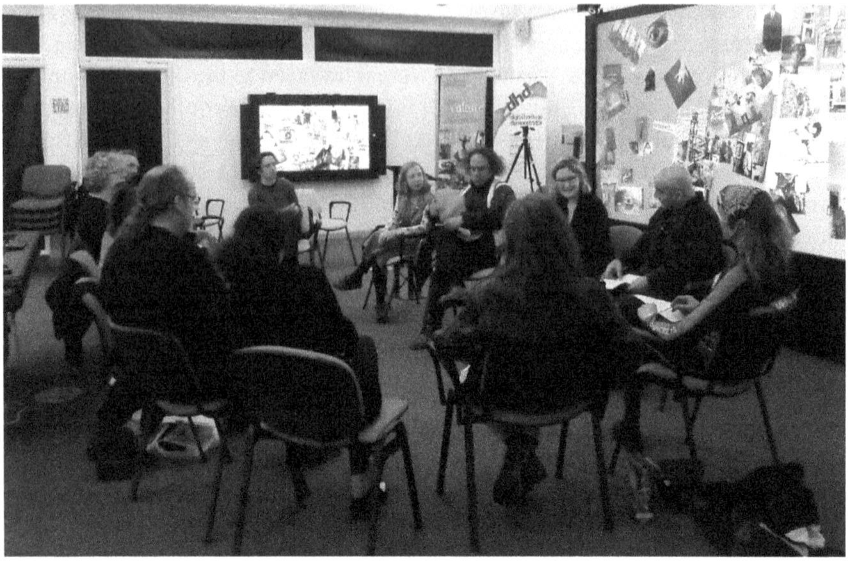

Figure 11.1 Playing *Exquisite Corpse* word games with workshop participants in the Birmingham Surrealist Laboratory, Digital Humanities Hub, University of Birmingham. Source: Saskia Warren and Stephen Forcer.

were introduced to Surrealism for the first time. In some of the participant-led collages, images were overlaid and juxtaposed in order to make an intended statement. For instance, one man placed an image of Margaret Thatcher head first into a toilet. This invited friendly contestation from another in the group that some Balsall Heath residents could be offended by the collage given support for the Conservative party fostered by David Cameron's recent visit to the area. Indeed, as the second participant observed, one local curry house proudly displays an image of the Prime Minister, who ate a meal in the restaurant on his birthday during the Conservative Party Conference in 2012.

While some collages were actioned deliberately, erroneous use of digital technology also created unintended surrealist collages of images. As Galloway argues, this kind of glitch can be seen as friction that is inherent to interfaces: 'an autonomous zone of interaction . . . concerned as much with unworkability and obfuscation as with connectivity and transparency' (2010: 120). One such result produced, as a participant termed it, the 'start of spaghetti junction', with all images clustering together in a swirling pattern that recalled local geographical points of reference. More studious arrangements were questioned: 'What does this mean to you?' 'I'm just doing it'. 'Is the point that there is no point? 'Yes, I wasn't really thinking about it.' This exchange recalls Breton's ideas on the role of unconscious thought in the creation of the artwork, where he argued that, 'Any form of expression in which automatism does not at least advance under cover

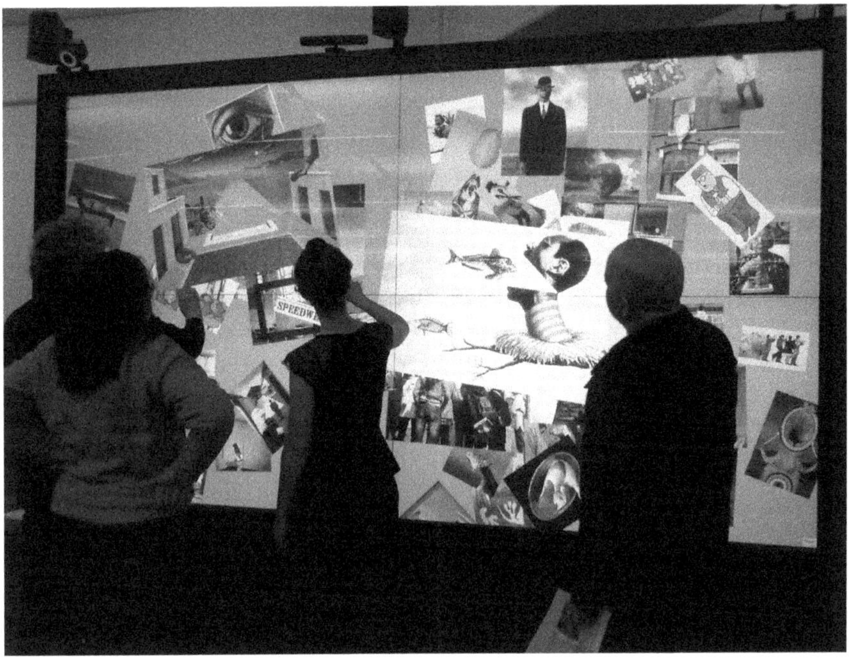

Figure 11.2 Workshop participants create Surrealist collages from digital interfaces, Digital Humanities Hub, University of Birmingham. Source: Saskia Warren and Stephen Forcer.

runs a grave risk of moving outside of the surrealist orbit' (Breton in Polizzotti 2002: xix). Or alternatively framed, a surrealist work 'must explore a mental space outside the field of normal awareness, in a place where consciousness and unconsciousness, the possible and impossible, become one' (Ibid. xix).

Workshop themes

The sections that follow discuss the key themes that arose in the workshop discussions: 1. Everyday Surrealism; 2. Cultural Heritage and Place-Marketing; 3. Imagination and Inclusion; and 4. Visibility and Offence.

1. Everyday surrealism

Recent historical events in Balsall Heath were considered fertile material for drawing out its surrealist credentials. In particular, allusion was given to the famous tornado of 2005 that ripped the roof tiles off a number of the terraced houses on Birchall Road. As one participant recalled:

> My friend's roof was lifted off of the house, and the way he can prove it is that his shirt sleeve was sucked up and caught under the roof of the house. He has photographic evidence. There were also pieces of roofing in the sky that looked like Pterodactyls. So there is natural surrealism in Balsall Heath.

Other incongruous events were relayed that pushed the boundaries of credibility. A participant talked about an occasion where a lion had escaped from a small zoo on Moseley Road, stressing: 'It had really happened.'

The educational currency of performing and teaching the history of Birmingham Surrealism within schools in the city was considered by one participant, a local teacher. She suggested that there is a surrealist theme in the annual Balsall Heath Carnival in which schoolchildren take part in a parade, as well as the potential for weaving it through the art syllabus: 'In the schools, they are doing GCSE and it [the work produced] is amazing. The children could learn Birmingham Surrealism.'

2. Cultural heritage and place-marketing

The cultural value of the Birmingham Surrealist Movement and its heritage was in the main interpreted in ways that aligned with commercial exploitation that was markedly distinct from a radically leftist communist or socialist approach often attributed to formal Surrealist practices (see Fenton 2004). How the heritage could be utilised in the area to support the economy was raised by a number of participants. As one said:

> I came [to the workshop] because culture and art for me are like food, really. Especially in Balsall Heath, there is a lot of creativity there. When you are regenerating an area it needs to have some kind of uniqueness to actually sell. I'm looking at it in a number of ways of how to make that Balsall Heath message a little bit different and unique.

Surrealism, although often connected with communism and socialism as a rejection of perceived rational, fascist and individualistic ways of organising society (Bonnett 1992, Fenton 2004), has always had a relationship with commerce. Salvador Dalí notoriously capitalised on the work he produced, generating great personal wealth in his lifetime, and advertising culture has widely appropriated Surrealist tropes and images (Mostafa 2005). This is significant for consideration given place-marketing by the participants with emphasis on the 'message' communicated of the area. Connecting the past with the needs of the present, and its changing demographic, was alluded to by another participant: 'If you look at the make-up of the area, what kind of message do you want to give out? There's quite a lot of responsibility there.' Instead of creating a barrier, however, the respondent felt that the social and cultural breadth of the demographic could be reflected in the ways in which the legacy of Birmingham Surrealists was disseminated:

> I think there are a lot of wonderful ideas, like a pop up where you can get people to express different ideas. If it is about education, and you want to draw a link to the historical or heritage aspect of Balsall Heath, then it needs to be explained.

Widening knowledge and understanding of the past is recognised as central to utilising the heritage of the area effectively, for adults and schoolchildren. For as one long-term resident of Balsall Heath acknowledged:

> My impression of Balsall Heath is it is all Joseph Chamberlain [former Prime Minister and Mayor of Birmingham 1873–6]. I don't know much about 1950s, the modern side. I haven't been taught about that.

3. Imagination and inclusion

Furthering education in the cultural heritage of Balsall Heath was considered important to fostering civic pride in the neighbourhood, which was part of a wider theme of inclusivity. As one participant adeptly observed: 'You don't have to have an art historical knowledge; it is about bringing your imagination so anyone can be involved.'

Beyond recognition given towards the inherent democracy of work centred on consciousness, including dreams and desires, was the appeal of its multifarious creative manifestations. This was particularly apparent through the way digital could be used to engage new learners with surrealist work and concepts:

> It is such an exciting project, it fires people's imaginations and what I think people would particularly like about it is that there are no rules. There's no right or wrong. That is the bit I would catch on to. It appeals to so many diverse groups and age groups, and it is something that interweaves across groups. Especially with digital concepts. No one has to be good at drawing. Such as collages, it could appeal to so many people. I can think of loads of people that would like to be involved.

More widely art was considered beneficial to well-being: 'Art is often something people go to in order to help themselves psychologically; [it] can cross boundaries and appeal to everyone.' This sense of border-crossing chimed with a number of participants, with one emphasising its potential to travel symbolically across social difference: 'It translates to different communities – and cuts across different groups.' The 'absence of pictorial cohesion or semantic coherence', in surrealist work, 'disturbs the viewing or reading subject' with 'information seemingly withheld' that 'frustrates and thus implicates the addressee' (Adamowicz 1998: 21). It requires 'cultural competence in decoding' symbolism and at the same time reveals 'irreducible heterogeneity' (1998: 25, 23). And yet despite the demands of the aesthetic, as is indicated above the democratic basis for understanding Surrealist ideas was part of its appeal to the participants. As one female participant suggested, it could potentially be used with vulnerable groups as a stimuli for talking about problems they are facing: 'It must be about young people; we need to involve them. I think about art as a means of therapy. People who are suffering from pain, or are refugees.'

4. Visibility and offence

Geographical proximity to cultural events and artwork does not define geometries of access, including whether artwork will be noticed, or whether subversive artwork will create an affect. Even public installation does not necessarily attract a general public, or engage those who pass by on a day-to-day routine (Zebracki 2013). Despite living and working in Balsall Heath, most of the participants at the workshops were unaware of recent cultural events that had taken place in the local residential streets and parks. As aforementioned, the *Balsall Heath Biennale* aimed to engage residents over a period of two years; a programme that included the Surrealist House competition where residents were asked to make and display a piece of Surrealist art in their homes and gardens. And yet only one participant knew of the Surrealist House competition (taking part and winning a prize for his efforts). As he noted, the Balsall Heath Biennale was widely advertised via 'quite a subversive newspaper as well', which was posted through the letter-boxes of houses and businesses across the neighbourhood. Another participant reflected:

> I have been working in the area for 17 to 18 years. I'm gutted I didn't know about the surrealist art on the houses [Surrealist House competition]. I would have had a lobster on the window at work.

The competition and newspaper represented what Adamowicz describes as the distinction between surrealism with a small or capital 'S': these were influenced by ideas of the formal Surrealist movement, with the potential to transform the everyday into the unexpected. While unfamiliar with the precedent of the Surrealist House competition, a participant noted, 'I would have come anyway. It seems like a great idea.' Little was known about the Birmingham Surrealist Group by the participants prior to the workshops, but the significance of the heritage for the area was keenly debated along with its changing cultural value over time.

Those who chose to participate in the workshops were a self-selecting community of interest that were open – at least in the context of the Digital Humanities Hub – to engaging and discussing the cultural value of Birmingham Surrealism. In this setting a lack of overt contestation over the sexualised and anti-clerical imagery of some of the images on the touch-tables was perhaps to be expected. A focus group forum for the discussion can also mean that participants are more likely to agree with the consensus rather than articulate dissent and, in doing so, avoid making themselves and their points of view more conspicuous. In the workshops humour around modesty was used by those who self-identified as practising Muslims as a means to deflect attention on their perceptions of the images. For instance, 'I'm not easily discomforted as long as we can keep our clothes on.' This exclamation fits a distinctly British tradition of slapstick humour and evokes the experimental Surrealist poet Gascoyne's line that the British are 'as scared of surrealism as they are of sex'.

But the *potential* for the form to create offence due to its sometimes sexualised imagery was implied in discussion on whether or not Surrealist imagery should appear in public space:

> Would it be inside four walls...? I'm open-minded. I'd go and see what it involves. But if it was on the billboards, if there was semi-nakedness, then people would get offended. If it's private inside then I don't think anyone would be offended. [It is an] issue of marketing.

The participant told us he was representing the views of his mosque. He introduced a pervasive distinction between public and private space in maintaining social order of civility in an area of distinct social, ethnic and inter-religious relations. Instead of a project or series of projects that cross-cut these differences, as suggested by other participants, it is politely proposed any work 'be inside four walls'. Following this rationale, convivial multi-cultural relations are supported when a shared cultural heritage, at least in terms of a community of interest around a formal area, are kept 'private inside'. A 'cartography of togetherness' where 'all kinds of people . . . take time, get affected by the environment and co-create the space and situation' (Nowicka and Vertovec 2013: 7) is viewed neither as something that can be controlled, nor uniformly desirable.

Gilroy (2004) argues, as outlined by Nowicka and Vertovec (2014: 4), that convivial culture 'renders racial and ethnic differences unremarkable: they become ordinary'. Instead people 'discover that the things which really divide them are much more profound: taste, lifestyles, leisure preferences' (Gilroy 2004: 39–40). This instance, however, suggests that while communities are marked by change and contingency rather than categorical fixity based around ethnicity, race and/or religion, axes of difference shape processes of identification with/in cultural heritage, and what is deemed 'appropriate' in public space. It is through taste, lifestyle and leisure preferences that racial, ethnic and religious differences can come to the fore. Conroy Maddox and the Birmingham Surrealists in the context of contemporary Balsall Heath highlight plural social norms that co-exist in the neighbourhood for which anti-religious and anti-clerical art might unsettle the ground, especially if forced into visibility and debate.

Re-visiting Surrealism in Balsall Heath: barriers and boundary-crossing

While geographical proximity is not the sole determining factor in widening public engagement, as the Balsall Heath Biennale demonstrates, accessibility is important. This point was made by one participant who argued that a workshop aimed at residents and workers in Balsall Heath should also take place locally:

> My feeling is that this workshop should really be happening in Balsall Heath, then more people would drop by. Such as the Old Print Works [a converted print company that now operates as artist studios and community space], where there are lots of artists who are a bit surreal, and they can drop in. It could happen on whatever scale such as Old Print Works, Ort [café and gallery], involve the Balsall Heath Forum and then merge into the carnival.

This narrative emphasises the use of existing infrastructure within the neighbourhood of Balsall Heath as a means of engaging local people through arts venues, events and community forums. Alternatively, another participant observed the potential of digital to break down physical boundaries and move away from a cultural asset-based approach:

> Would it [a Surrealist museum] have to be a real house? You have already demonstrated the potential with digital. You could build a virtual house which would have a hell of a lot more going on than in a material space.

Perceptual barriers to engaging with cutting-edge digital technologies remain. What this participant suggests, however, is considering the ways in which Birmingham Surrealism can translate to different media and different places. In fact, using the profile and facilities of renowned museums and galleries in the city centre or at the University of Birmingham could generate more benefits that extend beyond the formal boundaries of Balsall Heath:

> If we can we could take the artworks out, have it at the Barber Institute to give it value. If work is shown at Birmingham Museum and Art Gallery then it gives people pride.

Ultimately, the centrality of Maddox to an appreciation of the significance of Birmingham Surrealism and its heritage was stressed. Maddox is positioned as a maverick and change-maker who could inspire reflection on humanism:

> I think Maddox should play a big part in this. He has fired this talk, he shouldn't be left behind. He was influential at the time, and the reason we are sat here today is because of him. When you look at surrealist art it is about dreams and thoughts and the imagination and the ideas both inside and out. In a way that even the person who is doing it doesn't necessarily understand, it is simply pondered upon. People's imaginations aren't fired enough these days. Often people don't even notice a lot of things that are out there even. We need to draw attention to that past.

Conclusion

The Birmingham Surrealist Laboratory project investigates how the heritage of Surrealism travels and translates to residents who live and work in Balsall

Heath today. The contemporary nature of community and culture in a digital age are investigated through workshops that introduced innovative multi-touch tables and challenging forms of art in a super-diverse city. It worked directly with residents and workers from wide-ranging social, ethnic and religious backgrounds to open a cross-disciplinary and cross-sectoral dialogue that has significant implications for cultural and urban policy, arts industry and practitioners and academia. The mobilisation of cultural heritage for multiple strategies – education, civic pride, social inclusion, place-marketing – showed the ways in which communities of interest could form across axes of difference with overlapping but not always shared points of view. Through revisiting Surrealism in Balsall Heath the potential of building stronger inter-cultural communities around heritage has emerged, which could also inform strategies for growing the visitor economy in lesser-privileged neighourhoods beyond the city core.

An issue of visibility is raised even when working in focused ways in one neighbourhood over a duration of time. There are selected sight-lines meaning communities appear to remain unaware of events and resources available at the local level, even when efforts are taken to advertise widely through multiple channels. While the importance of physical accessibility to publicly funded events and resources is not in doubt, a shared geographical locale does not exclusively determine axes of inclusion. Instead this project points towards the salience of communities of interest for understanding the divisions in participation within local cultural and heritage programming. Moving away from specialised art historical knowledge of formal Surrealism, the findings instead championed understanding of the democratic humanism of the movement where commonalities are brought to the fore. This includes the ways in which people universally dream and imagine, and have desires and fears. Surrealism lends itself to play and even humour through the juxtaposition of seemingly unrelated concepts, which is part of its broad historical and contemporary appeal. In fact, the research foregrounds how Surrealism, understood as a way of thinking and doing, can be part of the everyday, but in ways that complement and expand the traditional focus on Paris as a privileged site of everyday Surrealist activity (Sheringham 2006). Furthermore, the role of art in contributing to understanding of psychology, therapy and well-being in community settings is highlighted by the case study as a valuable avenue of additional research.

Inter-cultural values around the heritage of Birmingham surrealism are challenged by some of the empirical findings in the chapter where it is suggested that the work is not publicly displayed to avoid the potential for offence. The issue is raised of whether conviviality in a multi-cultural environment is maintained most effectively, and perhaps counter-intuitively, by spatialised segregation of culturally and religiously subversive work into private space. Global movements of Surrealism, including in France, Egypt, Japan, Mexico, the Carribean and Scandinavia, are indicative of communities of interest that cross geographical borders. In 'unsettling the ground', the revisiting of the cultural heritage of Birmingham Surrealism within its wider global Surrealism context actually serves to raise important debate about the privileged role of art to challenge, to provoke and to offer alternative perspectives from the mainstream, accepted and familiar; a debate that intersects cultural and religious boundaries. At the very least, this project offers a starting point that indicates how communities of interest comprised of people from a broad range of countries of origins and migration trajectories, such as in Balsall Heath, could harness cultural heritage to boost place-marketing in their neighbourhood.

Further, and finally, this chapter contributes to the reevaluation of the 'continuing hegemony' of approaches that adhere to a close textual reading of cultural artefacts as stable cultural objects. As Rose (2015: 5) observes, in the past 30 years the cultural objects, technologies and modes of practice in society have significantly changed, along with processes of cultural production, display, participation and forms of critique. The Birmingham Surrealist Laboratory is responsive to the charge that too little has been said about how digital technologies perform roles in the creation of meaningful places. It sought to position resident groups as co-producers of the cultural heritage of the place in which they live and work, and its legacies. Dominant representations of an under-recognised neighbourhood and its residents are subverted, revealing Surrealism as a tool that unlocks hidden depths and histories, the unexpected and uncanny, tensions and multiplicities, and alternative visions and futures.

Acknowledgements

Many thanks to the Community & Culture Network+ who funded the research. We offer thanks to Professor Roger Shannon who participated in the roundtable discussion at the Barber Institute, Birmingham, along with Lee Saunders, Izzy Mohammed and Professor Elza Adamowicz. Additionally we offer thanks to Professor Richard Clay for his enthusiasm in the project and for usage of the Digital Heritage Hub, University of Birmingham. Finally, and most of all, sincere thanks to the participants in the Birmingham Surrealist Laboratory for their engagement and vision.

References

Adamowicz, E. 1998. *Surrealist Collage in Text and Image*. Cambridge: Cambridge University Press.

Adamowicz, E. 2014. Conroy Maddox and British Surrealism. Unpublished keynote paper given 5 July 2014 at the Birmingham Surrealist Laboratory. Birmingham: Institute of Fine Arts.

Ash, J. 2015. *The Interface Envelope*. London: Bloomsbury.

Barber Institute. 2014. Conroy Maddox and Birmingham Surrealism. Roundtable [Online: *YouTube.com*]. Available at: www.youtube.com/watch?v=fteByYP3x_Y [accessed: 8 August 2016].

Bonnett, A. 1992. Art, ideology, and everyday space: subversive tendencies from Dada to postmodernism. *Environment and Planning D: Society and Space*, 10, 69–86.

Breton, A. 2002 [1928]. *Surrealism and Painting*. Trans. Mark Polizzotti. Boston, MA: MFA Publications.

Elwood, S. and Leszczynski, A. 2013. New spatial media, new knowledge politics. *Transactions of the Institute of British Geographers*, 38, 544–559.

Fenton, J. 2004. 'A World where Action is the Sister of Dream': Surrealism and anti-capitalism in contemporary Paris. *Antipode*, 36, 942–962.

Forcer, S. 2013. *Conroy Maddox* [Online: *Vimeo.com*]. Available at: https://vimeo.com/68999438 [accessed: 8 August 2016].

Galloway, A. 2010. Networks, in *Critical Terms for Media Studies*, edited by W. Mitchell and M. Hansen. Chicago, IL: University of Chicago Press, 280–296.

Gilroy, P. 2004. *After Empire: Melancholia or Convivial Culture?* New York: Routledge.

International Surrealist Exhibition Catalogue. 1936. London: Women's Printing Society Ltd.

Leib, J. 1998. Communities of interest and minority districting after Miller v. Johnson. *Political Geography*, 17(6), 683–699.

Levy, S. 2003. *The Scandalous Eye: The Surrealism of Conroy Maddox*. Liverpool: Liverpool University Press.

Massey, D. 1994. A global sense of place, in *Space, Place and Gender*. Minneapolis, MN: University of Minnesota Press, 146–156.

Meissner, S. and Vertovec, S. 2015. Comparing super-diversity. *Ethnic and Racial Studies*, 38(4), 541–555.

Mostafa, M. 2005. An experimental investigation of the Egyptian consumers' attitudes towards Surrealism in advertising. *International Journal of Consumer Studies*, 29(3), 216–231.

Nowicka, M. and Vertovec, S. 2014. Introduction. Comparing convivialities: dreams and realities of living-with-difference. *European Journal of Cultural Studies*, 17(4), 341–356.

Polizzotti, M. 2002. Introduction, in *Surrealism and Painting*, edited by A. Breton. Boston, MA: MFA Publications, xvii–1.

Raco, M. and Flint, J. 2001. Communities, places and institutional relations: assessing the role of area-based community representation in local governance. *Political Geography*, 20(5), 585–612.

Remy, M. 2001. *Surrealism in Britain*. London: Lund Humphries Publishers Ltd.

Rose, G. 2015. Rethinking the geographies of cultural 'objects' through digital technologies: interface, network and friction. *Progress in Human Geography*, 40(3), 334–351.

Sheringham, M. 2006. *Everyday Life: Theories and Practices from Surrealism to the Present*. Oxford: Oxford University Press.

Sidey, T. 2000. *Surrealism in Birmingham 1935–1954*. Birmingham: Birmingham Museum and Art Galleries.

Vertovec, S. 2010. *Towards Post-Multiculturalism? Changing Communities, Conditions and Contexts of Diversity*. Paris: UNESCO.

Warren, S. and\ Jones, P. 2015. Local governance, disadvantaged communities and cultural intermediation in the creative urban economy. *Environment and Planning C: Government and Policy*, 33(6), 1738–1752.

Zebracki, M. 2013. Beyond public artopia: public art as perceived by its publics. *GeoJournal*, 78(2), 303–317.

12 A cybergeography of public art encounter

The case of *Rubber Duck*

Martin Zebracki

Introduction

Scholarship has largely focused on social engagement with public art in on-site material contexts, despite the proliferation of public art's digital dimensions (Freeman and Sheller 2015). Kellerman (2014) indicated that cyberspace is an emerging significant 'second action space' and aspect of everyday lived experience. Various strands of literature use different terms, including *cyberspace*, *virtual sphere* and *online space*, to capture contemporary conditions of interconnected global, abstract network spaces of computers and locative devices that enable both people and things to connect online (see Dodge and Kitchin 2001). The digital turn (Hartley 2012) has indicated the hybrid nature of the relationships between culture, digital technologies and society, where 'old' and 'new' media do not fit a historically linear pattern but co-emerge and merge in space and time – described as convergence culture by Jenkins (2008). Throughout this chapter I adhere to this convergence idea in exploring cyberspace through public art and what online public art does to people and places following on a holistic, hybrid understanding of its social and spatial publicness, temporalities and uses.

Public art studies have yet to more firmly frame geographies of online engagement with public art, particularly from bottom-up perspectives. By attending to people's everyday encounters with public art online, this article particularly answers Rose's (2015) call for more human geographical attention to participatory aspects of digitally mediated cultural objects. This article is based on a case study about the globetrotting *Rubber Duck* exhibition that has received substantial attention over (digital) media platforms since its first incarnation in 2012. I challenge prevailing understandings of public art as associated with the material world by analysing engagement along reconfigured comprehensions of spatialities and temporalities of public art's uses as manifested in cyberspace.

Dutch artist Florentijn Hofman, who has a considerable track record in large-scale urban installations,[1] drew inspiration for *Rubber Duck* from the Tolo-patented yellow bath toy. Since 2012, several inflatables, ranging from 5 to 18 metres in height, have been installed in about 20 urban waters across the world, including Amsterdam, Auckland, Baku, Beijing, Hasselt, Hong Kong, Kaohsiung (Figure 12.1), Keelung, Onomichi, Osaka, Pittsburgh, Saint-Nazaire, São Paulo, Sydney and Taoyuan.[2] Hofman's central mission statement reads:

> The *Rubber Duck* knows no frontiers, it doesn't discriminate people and doesn't have a political connotation. The friendly, floating *Rubber Duck* has healing properties: it can relieve global tensions as well as define them. The *Rubber Duck* is soft, friendly and suitable for all ages![3]

This quote suggests the artwork's capacity to bring different people and cultures together, trigger intercultural dialogue, reveal worldly social issues and help to mitigate those issues (while remaining 'apolitical'), and hence to have place-specific relevance. Considering *Rubber Duck*'s viral online coverage, I critically engage with the artist's rationale and global scale of ambition in the highly relevant but under-studied context of digitally networked space, which exists precisely through making connections with people and places around the world. In so doing, I throw light on the artwork's social significance and place-specificness as crystallised within digitally mediated engagement.

First, I analyse my central notion of the 'virtual relationality' of encounters with public art by dovetailing human geographical, new media and public art scholarship. The under-examined spatial and temporal dimensions of online (mis)uses of public art, alongside new possibilities for and critiques of online engagement are discussed, respectively. I then move on to the study's methodology. The results are presented and analysed along *Rubber Duck*'s degrees of social significance and place-specificness, followed by concluding reflections.

Virtual relationality

Scholarship has under-addressed the online roles of public artwork in everyday life. Starting from the premise that engagements in the offline world as well as online space should be seen in a dialectical and hybrid interface, I discuss three interrelated areas that support more firmly including cyberspace in public art scholarship. The first area is public art's spatiality: online engagements with public art have rendered conventional binary spatial notions (e.g. public/private and material/digital spaces) more fluid and less certain, creating novel ways for public exchange of experiences, and thus establishing a new under-explored sense of 'publicness'. The second area is a concern with time: digitally mediated public art engagements have created new understandings of permanence and ephemerality that do not necessarily tally with unidimensional, chronological temporal frameworks. The third area attends to the online (spatio-temporally) reconfigured *uses* of public art in a global hybrid space.

In their seminal geographical public art study, Massey and Rose (2003) introduced the notion of 'social relationality', grounded in Lefebvre's (1991) theorem that space entails a socially *co*-produced entity. 'Social relationality' is defined as the very essence, and therefore 'publicness', of public art: 'for an artwork to be public [and hence meaningful], negotiation between social differences has to be part of what the artwork does. If negotiation among diverse social identities is not invited, then the artwork is not public' (Massey and Rose 2003: 19). They applied their notion to public art's potential uses and meanings as situated in a material world,

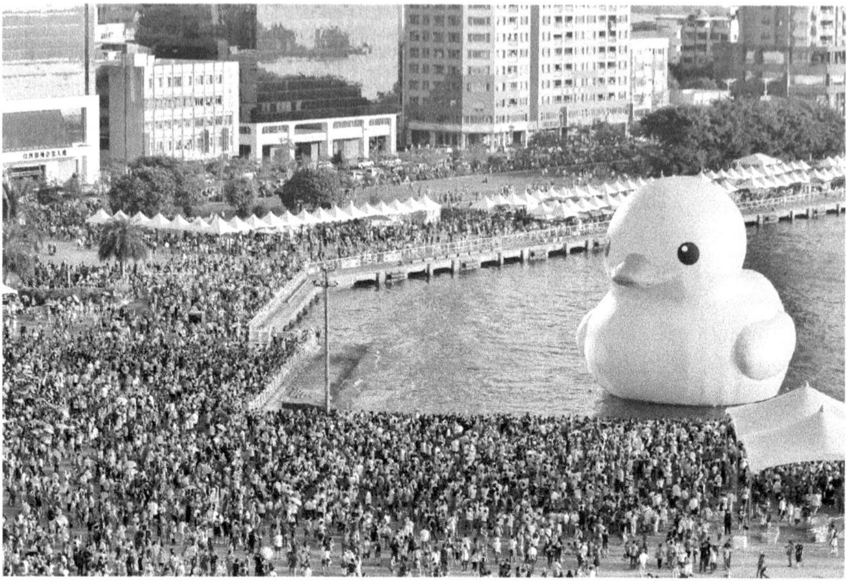

Figure 12.1 *Rubber Duck* (18m), Kaohsiung, Taiwan, 2013. Source: Courtesy of Studio Florentijn Hofman.

which strongly depend on the artwork's physical, spatial and temporal properties (e.g. size, colour, shape, position, permanence/ephemerality). Not all people engage with the same gamut of properties; also, they pursue these properties to different degrees and in different ways, making any generalisations impracticable (Massey and Rose 2003: 19).

I extend the notion of social relationality to what I term 'virtual relationality': the mediation and appropriation/repurposing/challenging of public art's properties and roles in digitally mediated social relations and hence networked spaces. Thus, the focus is on online users' social (re)constructions of public art in digital communities, covering popular social media outlets focused on user-created content (i.e. Web 2.0). These include Twitter, Facebook and YouTube, as well as forums and online news platforms, which differ in user-creation potential and bottom-up social agency (Hartley 2012).

Online uses: reconfiguring spatialities and temporalities

Virtually mediated public art requires a reconfigured understanding of space and time vis-à-vis public art engagement prior to the emergence of digital culture. Social media have substantially affected and digitally intensified quotidian life (Brake 2014, Hartley 2012). Although digitally mediated content as such does not carry any spatial-material 'mass' for tactile engagement (see Dodge and Kitchin 2001), it is layered over the material world and people's 'real-world' experiences. As argued by Crang et al. (1999: 1): '[digital] technologies are seen as facilitating, if not producing, a qualitatively different human experience of dwelling in the world; new articulations of near and far, present and absent, body and technology, self and environment.'

Where offline public art encounters typically involve multiple in-situ users within single material environments, online and particularly digital mobile public art engagements may involve digital multi-user environments to overlap with multiple material environments (see De Souza e Silva 2004). Such 'messiness' of space in digital culture (Rose 2015) fundamentally reconfigures the dominant Habermasian notion of the public sphere: contemporary digital culture is spatialised along fluidly dispersed, and potentially invisible, publics (Hartley 2012). Social relations and identities are, accordingly, expanded and reconfigured by multi-directional technological-digital practices such as uploading, downloading and streaming through desktops, mobile devices and do-it-yourself (DIY) technologies (Hartley 2012, Kidd 2014). This fits Jenkins's (2008) understanding of convergence culture, wherein offline/online engagements and analogue/digital and hardware/software technologies coexist in 'real life' and real time, thereby augmenting one another and deconstructing open/closed space divides.

The 'digitisation of our existence' (Bishop 2012: 436) has further blurred public/private and physical/immaterial boundaries, and provided augmented and immersive understandings of public values of cultural objects. Also, it has provided the empowering opportunity for everyday citizens to create content beyond institutionalised spaces; for example, through YouTube, Instagram, Tumblr (Kidd 2014). This essentially marks the shift from people as passive audiences/visitors/end users towards participatory, self-authoritative, purposive and creative *publics* (Goriunova 2013, Kidd 2014), consequently boosting grassroots agency, collaborative meaning-making and collective intelligence (Hartley 2012). Hartley (2012) argued that social media in particular (e.g. Twitter, Facebook, YouTube, Tumblr) offer fewer limitations to participation and interaction than more conventional media that have been extrapolated into the digital sphere, such as online television.

Digital surfing can involve myriad uses of art: for example, browsing, sharing, creating, editing and curating or composing first- or second-hand art content, which can be done over, for example, image management and sharing platforms of Flickr and Pinterest (Kidd 2014). Digitally networked space enhances the affective experience of public art (Gauthier 2015), as it essentially inscribes (and networks) publics around the artwork's appreciation through *ekphrasis* (discursive commentary) (Rhodes 2015). This expands Warner's (2002: 62) definition of publics to the digital: 'a public is understood to be an ongoing space of encounter for discourse. It is not texts themselves that create publics, but the concatenation of texts through time.' Rather, as textual and visual elements of user-created content are often digitally amalgamated over digital sites and time, I would prefer to speak of dynamic/moving 'text-images' (e.g. emoticons, annotated photos). Rhodes (2015) provides the example of proliferating augmented reality applications for mobile phone users to visually and discursively 're-texture' real-world environments, including public streets and museums (e.g. the Louvre and the Guggenheim). So, after Boellstorff (2008), online activities can inform publics' expressions and 'doings' in the public artwork's material locality,[4] and the other way round – producing temporally dynamic social, spatial, material and digital parameters of public art experience.

Online interactions can occur in a synchronous, real-time manner (e.g. live tweeting, interactive Facebook chats) or in an asynchronous fashion: users can communicate, at different and self-selected times, about public art through, for example, forums, weblogs, social media profiles. After Dodge and Kitchin (2001), digital public art content can instantly emerge and disappear 'in a click', denoting an ambiguous status of being permanent, temporary or somewhere in between. For example, a temporary artwork removed from its physical locality might 'live on' as entity on the Internet or be 'resurrected' through a mobile augmented reality application.

New possibilities for engagement

Digitally networked relations can put local and global 'whole populations' in contact over vast geographical distances (Hartley 2012). Publics across the global digital world can be connected around an artwork in a specific material locality (if any at all) and interactions can be particularly intensified on platforms with large communities (e.g. Facebook and Twitter). In comparison to offline interactions, it should be noted that online users sometimes have more substantial control over (dis)closing identity: for example, 'real' profile photo vs avatar on Facebook; active tweeter vs passive 'liker' or 'silent voyeur' (see Kidd 2014). Moreover, although digital participation might be commonplace, it is not necessarily possible everywhere and equal for all people in society, considering the requirements for Internet access, digital communication skills/literacy and membership of online accounts (see Brake 2014).

According to Stevens and Lossau (2015), the making and consumption of public art have traditionally been strongly conditional on authoritative control and the 'privileged category of engagement' of sponsors and artists (2015: 7), who impose *functions* and 'manipulate audiences so that they will perform what the artwork prescribes' (2015: 5). They employed the word *use* to address a power shift to publics, 'who find their own purposes in the aesthetic objects and experiences presented to them' (2015: 5). Applying their argument to digital culture, this power shift becomes particularly notable considering online users' potentially strong possibility to (re)imagine, (re)purpose and (re)occupy spaces through especially social media (which are usually user-friendly and free of charge) (Kidd 2014, Lodi 2014).

This grassroots condition of what can be considered 'post-Internet art' (Gauthier 2015) has incited a new digital consumer agency and democracy (Hartley 2012, Jenkins 2008) which breaks the art world's conventional artist/audience, expert/novice and authenticity/quality binaries. 'Amateurs' have gained digital capital to, for example, crowdsource and produce high-quality content to 'game' established media and art-world systems (Kidd 2014). So, digital publics might be considered real self-organising citizens whose content can be appreciated through the 'digital aesthetics' of bottom-up/DIY creative (co-)production. This produces novel digital knowledges and senses of digital citizenship and increases the legitimacy of ordinary views and grassroots engagements. This legitimacy, however, is often challenged by authorities' formal evaluations of 'expert' output and 'measurable' quality (Kidd 2014).

Critiques of online engagement

Why and how do publics engage with public art online? Miller (2008) argued that online users, depending on any cultural and linguistic communicative barriers, navigate through individualisation (e.g. personal stories/blogs with little social exchange) and network sociability: social bonding practices to construct and maintain personal digital networks (e.g. Facebook and Instagram accounts). Particularly in the context of network sociability, Hartley (2012) observed upsurges of play and humour, indicating acts of 'silly citizenship'. These imply whimsical, light e-communications (e.g. sharing 'cool'/'fun' pics and memes), which may subvert elite, seriously informed and adultist rationalities. Despite the light-hearted intent of such communications, Goriunova (2013) deploys an existential critique that reveals the dark matter of light e-communications: in the author's view, the *en masse* techno-human subjectification has involved a performance of idiocy revealing 'the trouble of the current human condition' (2013: 11).

Miller (2008) voiced a similar nihilistic assessment of much indiscriminate digital communication, which he termed phatic communication – 'communication without content has taken precedence' (2008: 398) – moving away from substantial discursive dialogue towards the primary purpose of conveying sociability and maintaining a social network. Miller (2008) associates phatic communication with database principles of endless growth (e.g. constant microblogging on Twitter); obliviousness of informational production/consumption (see Hartley 2012); prevailing 'promiscuity' in network sociality (e.g. ceaselessly adding 'friends'); the desire to be 'followed' or 'liked' on Twitter, Instagram, etc.; and expectations of ongoing 'connected presence' (Licoppe and Smoreda 2005). The latter denotes 'must have' feelings of being/staying connected with other online users through Internet and mobile devices everywhere and at any time – while there is a coexisting paradox of being present and absent.

Phatic communication might be imposed by the medium itself. Twitter allows 140 characters per tweet only, but social networks such as Facebook, Instagram and Tumblr do not impose any character limit, although a conditioned code on social networks is to keep messages as short as possible (Kidd 2014). Particularly the popular platform of Facebook, compared to the prevalent microblogging, text-focused platform of Twitter, offers substantial functional space and visual prominence for network sociality and thus maintaining social relationships (Kidd 2014).

Online practices can be motivated by, for example, (phatic) indifference and everyday banalities, or emotional attachment and ensuing heated debates, acts of public shaming, and trolling (i.e. posting inflammatory content in discussions over the Internet) (see Brake 2014). Amin's (2002) notion of 'micropublic', originally designating offline spaces for intercultural encounter, offers a relevant angle for exploring public art's social significance. A digital micropublic would, then, entail digitally mediated space, ranging from indifferent to antagonistic and judgemental to non-judgemental atmospheres, wherein publics encounter, exchange, negotiate, bridge, challenge and/or potentially transgress everyday social and cultural values, differences and norms through public artwork (after Bishop 2012).

Richard Serra's *Tilted Arc*, a long curved wall traversing NYC's Federal Plaza (1981), served as a classic example of how public resistance in the real world, as mediated by print media, radio and television, led to the destruction of this public artwork in 1989. The vandalisation of Paul McCarthy's *Tree*, a 24m-tall inflatable sculpture resembling a butt plug in Place Vendôme in Paris in 2014, demonstrated the power of social media in reciprocally affecting public art online and offline (see Zebracki, forthcoming). Gauthier (2015: 29–30) conveyed that the *Tree* case is a:

> noteworthy story of the rejection of a public artwork in a networked age . . . it revealed rifts in the fabric of the French public . . . the virtual layer outstripped the physical . . . the assault transferred from the online to the physical realm.

(The artist was even slapped in his face by an offended onlooker.) The heated debate was accelerated online: *Tree* became a meme and ludic visual reinterpretations circulated, which challenged heteronormative society and aristocratic architecture. Some even took this public artwork as a worthy successor of the Eiffel Tower. *Tree* is physically no longer in the Place Vendôme, but continues to exist as a conversation piece both in 'real life' and online. This example illustrates the creation of a digital micropublic, wherein users digitally mediated 'co-authorship' of public art and repurposed its roles along reconfigured senses of space and time.

Methodology

Considering its rich digital exposure since its first installation in 2012, *Rubber Duck* is a pertinent case for analysing online public art engagement. The study of this popular urban installation, a genre with which I am highly familiar, is not isolated and might produce inductive insights into online encounters with public art of similar calibre. I conducted 'virtual ethnography', described by Hine (2000: 8) as a useful method 'to develop an enriched sense of the meanings of the technology and the cultures which enable it and are enabled by it'. This is also a particularly useful method to reach and involve publics who might be difficult or impossible to approach in 'real life' (Kidd 2014). Following Driscoll and Gregg's (2010) proffered advice, I conducted e-observations in a covert manner so as not to intervene in authentic communicative flows.

Online discussion of *Rubber Duck* has mostly been taking place in the form of 'text-images' on the social media platforms of Twitter and Facebook. Annotated photos and videos have moreover been circulated, particularly over Pinterest and YouTube respectively. Artist statements and project descriptions circulated on the artist's studio homepage, international tourist boards and listing sites (to arouse interest in on-site visiting), and news platforms in relation to nearly every exhibit. About 100 unique, distinct items (created/published by formal agents as well as everyday online users) were gathered to gain a solid impression of *Rubber Duck*'s coverage in digitally networked space. This was largely done through manual sampling on search engines of social networking websites and Google – entailing a snowballing technique as content often referred to/hyperlinked other content. Distinct items were considered, among other things, posts, news stories, op-eds, videos and tweets – shares/crosspostings of second-hand content, retweets and intra-post comments (e.g. replies to YouTube films and comments on Facebook posts) were screened but not counted as such.

The content was usually a hybrid of textual and (audio)visual materials, mostly photographs (see 'remix culture' in Kidd 2014). Keywords and hashtags, referring to the artist, artwork and the artwork's various physical locations, yielded a geographical breadth of information. The collection involved one automatic retrieval method: the Twittersphere was also scraped with TweetDeck, enabling tweets to be located in a set radius (e.g. 5 km) around the artwork's GPS location (a form of 'geofencing').

Content accumulated and evolved over the duration of the *Rubber Duck* project, enabling an aggregated and diachronic textual and (audio)visual discourse analysis. I identified discursive threads in the artwork's online uses along spatial and temporal dimensions until I reached an unprompted stage of data saturation. This also involved thick description: the analytic embedment of online engagement with *Rubber Duck* in broader digital media contexts (wherein I am firmly situated as an active social media user). On ethical grounds to do with ownership/usage of user-created content (see Kidd 2014), only publicly accessible online scripts and expressions were analysed and are anonymously reported here. Accordingly, URLs of user-created content are not included in this chapter.

I complemented the virtual ethnography with an in-depth semi-structured expert interview (administered in Dutch) with *Rubber Duck*'s creator at his home studio (11 November 2014), who provided consent for identifying his views. The 'inter-view' implied a setting of trust allowing me to conversationally and viscerally situate my findings within the artist's frame of mind as it had evolved throughout the project. In a triangulating fashion, I post-analysed findings from the virtual observations, discourse analysis of secondary data and in-vivo/first-hand data analysis.

Results and discussion

Analysis of online uses of *Rubber Duck* revealed various degrees of engagements, ranging from dialogical to rather phatic communications that disclosed differing associations with time and space. Sharing, commenting and curating/editing practices were the most salient interrelated engagement types, which are hereinafter questioned regarding, first, the extent to which engagements can be grasped in the spirit of critical or somewhat phatic digital micropublics – to what degree do online users unleash and negotiate any critical thoughts and deep feelings (or not)? Second, engagements are analysed for the extent to which they unfold any everyday place-specific attachments or rather global, universal and thus more generic spatial and temporal connections.

Critical or phatic digital micropublics?

Rubber Duck's online coverage had accelerated over the project, and Facebook and Twitter in particular were strewn with user-created content. I observed that most of the content was intended to entertain both content-creator and content-consumer. Online promotional materials by Hofman's studio and local authorities hosting *Rubber Duck* on their waters emphasised the cultural grandeur of both the artwork and city. User-created content, again and again, particularly played along with the artwork's flagship allure and sensational spectacles of official unveilings.

A salient illustration was the massively attended and richly online documented entry of *Rubber Duck* into Keelung, Taiwan. The inflatable became even more virally reported on New Year's Eve 2013/14 when it 'exploded' in front of a large crowd, just 11 days after its installation. Dramatic news reports presented headlines like 'it's a Duck meltdown'.[5] Online speculative threads about the explosion's cause reinforced the sensation-seeking spectacle; for example, online users wondered whether this resulted from any technical issues or vandalism. Hofman (interview) refuted any dramatic hearsay and imparted that *Rubber Duck* was preventively deflated due to an approaching typhoon. Hofman loathed (social) media's ongoing pursuit of theatrical news value and parroting reporting at moments of 'crisis': 'they hear of something without knowing the origin', he said.

Rubber Duck's online coverage repeatedly entailed entertaining narratives characterised by random snapshots as embedded in everyday trivial activities (which bear a strong resemblance to the *Tree* case; see Zebracki, forthcoming). This suggested that the artwork carried meaning, no matter how subtle, in people's 'little daily things'. On Studio Florentijn Hofman's Facebook account, the artist occasionally shared user-created, banal imageries and stories about the work alongside news headlines and official promotional materials. A telling example was a Facebook photo titled *Enoying* [sic] *ice creams*. This image showed a senior couple eating ice creams in front of the installation, floating on Parramatta River in a public park during Sydney Festival in January 2014. As of 19 March 2014, 965 online users worldwide had clicked this photo's 'like' button, 22 had commented on it and 59 had shared the post. Most of the shares indicated uncommented, fairly facile referrals. These, in their turn, were often liked or commented upon by Facebook friends of the respective users, demonstrating vivid interactions. The comments were largely brief. Rather than environmental references, they stressed 'people-art' emotions in the spheres of aesthetics, amusement, sympathy and fun; for example, 'so cute', 'lovely', 'enjoying *Rubber Duck*', 'best pic', 'enjoyment of innocence and his wife',[6] and 'who let the duck out?' – an array of narrated interventions that may have been indicators of 'silly citizenship' (Hartley 2012).

I experienced this example as a quite personalised portrait of *Rubber Duck*. Online sharing appeared to interact between the intimate scale of the digital photograph, the artwork's material scale, and the Internet's global scale. This expanded the artwork's publicness in a multi-layered way: it showed how something personal had been made accessible to online users anywhere in the physical world. Moreover, the posted photo of *Rubber Duck* with the ice-cream eating couple can be rendered as a hybrid, 'more-than-human' ecology: it interlinked art matter with human corporeality and digitally mediated not only a sense of everydayness but also everywhereness. Ice-cream eating was made a familiar yet extraordinary experience by the artwork's presence. Through the digital text-image, users could transport themselves mentally into similar simple joys of everyday life, potentially providing vicarious experiences and moments of (virtually distant) interpersonal encounter.

Another salient example of an interactive Facebook thread (combining practices of sharing, visual editing and textual commenting), laid bare a playful yet critical eye for cultural consumer behaviour. With combined reference to Hofman's 'air sculpture' and Robert Pirsig's icon of American cultural-philosophical literature, *Zen and the Art of Motorcycle Maintenance: An Inquiry into Values*, a Facebook user shared one of Hofman's photographs of *Rubber Duck* and whimsically annotated: 'Zen and the art of eating soft serve' (i.e. ice cream made softer by the use of air). Another Facebook user entered into a dialogue and commented: 'where's the pitchfork?', to which the first commenter replied: 'American Gothic *Rubber Duck*. I can do that.' This provoked a reaction: the user of the original post uploaded an edited image of the much-parodied *American Gothic* (with 'apologies to Grant Wood'), depicting a yellow duck protruding between the farmer and spinster daughter. The other user amusingly commented: 'they look like they are about to duck.' The first commenter then wittily replied in reference to what has become *de rigeur* in social media: 'duck wants in on the selfie.' This user's comment can be critically linked to 'plop art' (Kwon 2004): perceived random (and potentially intrusive) sitings of public art around the world – just as some people have come to experience the act of taking selfies as inconsiderately invasive in (public) art spaces.[7] The latter is also indicative of phatic online communication, where 'the ephemerality of digital effort is always flickering with the energy of immortalised selfies' (Freeman and Sheller 2015: 6).

I consider the two previous examples significant illustrations of Dodge and Kitchin's (2001) argument on temporality in digitally networked space. The respective digital posts and associated threads appeared to be offline when I attempted to revisit them as of the time of writing, in June 2015. Philosophically, there is nevertheless 'connected presence' (Licoppe and Smoreda 2005) in that the current post-analysis resuscitates this content in a sense and makes the online users' encounters present again (albeit in their virtual absence).

Furthermore, I observed that a substantial number of encounters with *Rubber Duck* over social media lacked any deeper social critique in their discursive and visual semiotic contexts – resonating with Miller's (2008) theorem of hegemonic phatic e-communication in digital network sociality. The examined engagements seemed to be mostly situated in atmospheres of sensation and echoed everyone-pleasing and consensual art-making, as also insinuated by the artist's mission statement. Public news debate challenged the artist's statement, resulting in ample digital footage. For instance, in an online broadcast of the popular Dutch late-night talk show *Pauw & Witteman*, the interviewer Jeroen Pauw considered Hofman a shrewd entrepreneur who uses *Rubber Duck* as a popular mass article: 'if you make something big it will catch the eye, so people find it attractive to have the bath duck floating in their harbour.'[8] Hofman explained that he materially scaled up the yellow bath duck to create the feeling of a global, harmonious artwork. 'The worldly waters have become our bathtub', Hofman said, and therefore he believed that his artwork reconciles and breaks down all barriers between people. Hofman imparted that some citizens of Pittsburgh petitioned local authorities to prolong the artwork's presence: 'people seem to fall in love with it and want to keep it.'

That said, there was online activity about how *Rubber Duck* was used for geopolitical purposes, which uncovered social and political tensions that might mount around large-scale urban art installations like this one. The Chinese Nationalist Party (KMT) in Taiwan 'hijacked' the artwork as the figurehead for a presidential election campaign vibrantly played over social media. 'In *Rubber Duck* we would have a unique president who could even cross the Taiwan Strait by itself, depending on the prevailing wind', said a KMT representative.[9] Hofman (interview) asserted that over social media 'many lies were going around about alleged high sums that politicians offered in the race to get the bath duck to their location for political profit'. This suggested that city authorities, in China in particular, used *Rubber Duck* as an online plaything to underscore their importance within political hierarchies in 'real life'. Moreover, China's authoritarian government and its official control formed a critical motif in some online users' appropriations of *Rubber Duck*. Most notably, in 2013 a user of the popular Chinese microblogging and social networking site Sina Weibo posted a photoshopped version of the 'Tank Man' photograph close to the anniversary of the student-led Tiananmen Square protests on 4 June 1989. The edited image featured a column of *Rubber Ducks* instead of tanks and soon became a meme (with some quirky variants), as in the *Tree* case (see Zebracki, forthcoming). Some highly placed officials engaged with this meme. For example, while the Chinese government censored various words including 'big yellow duck' on Sina Weibo, a United States Foreign Service Officer tweeted 'Chinese netizens 1, Chinese censors 0'.[10]

Nonetheless, everyday online users' 'prosaic' and 'banal' acts (Amin 2002) primarily set the tone for *Rubber Duck*'s digital micropublic – rather than sustained dialogues and practices subverting the established order. Mostly by posting, commenting and curating/editing, publics offered other users text-images as digital 'contact zones' to react upon (after Askins and Pain 2011). Unilateral communications and banal interactions typified these micropublics. Examples were legion: 'if you still do the duck face in pictures I'm judging you' represents one of the many one-liners in the Twittersphere. Snapshots were tweeted of people kissing and squeezing *Rubber Duck*. The depicted poses drew a compelling parallel with tourists' photo-taking behaviours in offline space (e.g. clichéd snaps of people apparently holding up the Leaning Tower of Pisa). The previous snapshots sparked off interaction between online users and Hofman himself – a direct connection that might be unlikely to occur in offline space. Hofman retweeted the snapshots and added to the latter: 'don't squish me!'

Although this small intervention showed, in the vein of Amin (2002), a prosaic dialogue, it revealed interesting banal transgressions at two levels. First, Hofman, by his simple line, discursively incarnated his own artwork in online space. He augmented *Rubber Duck*'s offline postmodern visual language with a digitally transmitted ironic-textual dimension. Also, despite the artwork's super-large material scale, Hofman's irony hinted that *Rubber Duck* can be 'cuddled' online, thus making it more-than-visual and 'digitally tactile' within online experience, which contradicts the 'only-visual' existence of digital content as implied by Dodge and Kitchin (2001). Second, this example showed how the use of public artwork through social media can lower the threshold for everyday interactions between the artwork's creator and its audiences. Also, the Twitter thread under discussion might be described as an interactive interpretative panel thrown open to the virtual public at large.

Hofman (interview) moreover indicated that *Rubber Duck*'s online omnipresence might have made him more approachable as an 'emailing body', too. His art studio receives and replies to daily fan email. As we spoke, Hofman printed out and discussed an email that he just received from a South Korean schoolgirl. She enclosed her school essay about *Rubber Duck* and wrote that its 'cute appearance' inspired teachers, friends and herself. Hofman pointed out that the profuse e-communication about the artwork both revealed and contributed to real-world impacts on people. But it remained largely indistinct to me how social relationships were exactly deepened in digitally networked spaces.

Place-specific or universal?

Rubber Duck's generic style and touring nature raise questions about the degree to which this public artwork speaks to specific places and publics *or* rather to global (art) trends and global (digital) markets and audiences. Hofman's oeuvre resonates with the postmodern aesthetic of determinedly oversized and kitschy artwork by leading contemporary artists like Jeff Koons and Paul McCarthy – see analyses of McCarthy's 'Butt Plug Gnome' in Zebracki (2012) and *Tree* in Zebracki (forthcoming). Some might consequently experience this 'uncultivated' aesthetic as disconnected from local architecture and falling short of deliberate dialogue with local publics – see 'plop art', which according to Kwon (2004) reflects the globalised art world's marketed habits/desires, resulting in the 'commodification and serialisation of places' (2004: 55). Others might, nevertheless, experience deeper meanings, locational awareness and connectedness with local or global (digital) publics.

I observed an analogy between online engagement with *Rubber Duck* and its commodification in offline space. The travelling exhibition could, indeed, be considered a serial spatial project propagating homogeneous expectations about its object (enlarged plastic inflatable bath toy), location (urban water and skyline as backdrop), audience (large voyeuristic crowd) and habits (mostly photo-taking and merchandising practices). Digitally mediated images/imaginations reproduced and enhanced these expectations and, as such, amplified non-concrete social identities and space-times.

That said, some place-specific text-images of *Rubber Duck* were created through mobile devices on the spot (which could be seen as portable digital simulacra of the artwork). Online users occasionally carried over experiences while being among crowds beholding the artwork in situ, channelling feelings of encountering the artwork together. I encountered the content through my computer in my own geographical localities (university campus and home space), situated in places and times different from the exhibitions. The intimate scale of some online materials let me, in some capacity, vicariously co-sense mediated experiences of in-situ onlookers. For example, I gained a relayed experience from my encounter with this tweet: 'lol @ the giant duck floating around campus.' The Twitterer accompanied this post with a photograph taken on the artwork's site in real time. This instance vividly played on my imagination of what it would mean to experience *Rubber Duck* in my everyday locale (e.g. as intruding, aestheticising, alienating, place matching?).

Amateur entries were often interlinked with professional news narratives. As such, e-communications (a)synchronously linked to other digital places and offline realities, awaking an indeterminate, digitally compressed sense of space-time. Although Miller (2008) identified much social media activity as apathetic and indifferent, its viral potential could monopolise some explicit (positive or negative) attitudes and lend colour to public debates accordingly. Some digital encounters with *Rubber Duck* were outspoken. *CityLab*, for example, published a clearly critical op-ed, shared by more than 1,000 social media users:

> The *Rubber Duck* artist must be stopped: the inflatable spectacles of Florentijn Hofman don't belong in every harbour in the wide world . . . *Rubber Duck* sends an infantilising message about the role of public art in cities.[11]

The author applied similar critiques to Hofman's other travelling installations, including the floating supersized wooden hippo sculpture. Strikingly, content created by both official media and grassroots largely presented various *Rubber Duck* exhibition spaces as spatially and socially undifferentiated and disassociated. 'Unique' environmental elements, beyond the presence of water for the *Duck*, were not quite fleshed out through artwork properties and complexities. A broadly homogenised online gaze fetishised images of the bath toy and the object's large-scale stature in lieu of attending to the diversity within publics and what the artwork really does to the *couleur locale*. Although some online narration indicated place-specific experience, universal place images dominated. A thorough public induction by the exhibition organisers and promoters could have been an opportunity to heighten awareness of this artwork's potential site specificity (see Kwon 2004) on the part of both online and offline publics.

Nevertheless, spatial attachments occasionally clearly filtered through user-created content; for example, in the guise of references to other *Rubber Duck* exhibitions that called up its geographical trajectory. A Facebook thread about the Sydney-based installation included a user's post of a self-taken, annotated photograph of *Rubber Duck* in Pittsburgh: 'thank you [Hofman] for bringing your duck to Pittsburgh. We loved having him here!' Such gratitude/interest, no matter how trivial, might precisely be the public response that flagship-art marketing strategies aim for, both offline and online.

But sometimes goals were more ambitious. The Netherlands Embassy in Beijing invited Hofman to deliver a press conference as part of 2013 Beijing Design Week as well as Guest City Amsterdam.[12] This did not just serve to promote the spatial identity and creative economies of both cities and their respective countries. Many counterfeit goods appeared to circulate, particularly over mainland China, and therefore the Design Week official declared: 'We want to use the *Rubber Duck* case to drive an awareness programme raising the sensibility towards intellectual property rights around China.'[13]

The online incarnation of *Rubber Duck* suggested an enactment of globalised, commodified values of popular culture. Various offline as well as online pop-up stores mushroomed mainly in the Far East, and especially in South Korea and China, selling all kinds of products, including bath supplies, eco-bags, mugs, beddings, smartphone cases, T-shirts, school supplies and even humidifiers, with slogans such as 'enjoy the *Rubber Duck* and get energy from the duck in your daily life!' (Song 2014). Such (e-)commerce embedded in an economic culture of imitation might not only mutually reinforce phatic online behaviour (see Miller's (2008) grander critique of increasing nihilism in western (e-)culture). It might further market/standardise spaces for public art engagement and, thus, cultivate 'hands-off' relationships with specific offline places. Hofman, in our conversation, acknowledged being somewhat flattered by the various (counterfeit) pastiches. At the same time, he advanced a critique of neoliberal livelihood strategies: much (online) merchandise of *Rubber Duck*, especially within the scope of illegal trade culture, just indicates a way to scratch a living by merchants coping with broader, large-scale commercial competition.

Some online commenters criticised the artwork's dearth of unique spatial quality. In the comments thread of the YouTube tabloid video *Florentijn Hofman's Giant Rubber Duck to Swim to Hong Kong Harbour City*,[14] a user made a plea for an alternative public artwork: 'hmmmm might have been better if they did giant pink dolphins, the ones that are becoming endangered because of pollution in the Pearl River :o' (i.e. 'surprise' emoticon). In this video, Hofman conveyed that '[*Rubber Duck*] brings back memories of your childhood; and as everyone had a childhood, the *Rubber Duck* connects everyone.' A critical screening renders this statement as eschewing social differences, which thus challenges the artwork's 'publicness' following Massey and Rose's (2003) criterion of social relationality. Hofman (interview) pursued a romantic naïveté in an ambiguous way. Although he demonstrated awareness that people could take the artwork as universal/non-place-specific, he indicated its place-specific and supra-spatial/global relevance and potential to stir intercultural curiosity:

I am in duty bound to enter into engagements with the audience . . . *Rubber Duck* laboratises human behaviours . . . It is a catalyst: it shows what is going on and makes people around the world interested in new places.

Hofman (interview) imparted that he received a great number of positive testimonials from arts and civil society organisations, as well as individuals who enjoyed *Rubber Duck* through both analogue and digital communication channels. For example, he received a box of chocolates from the father of an autistic child for whom this artwork seemed to have therapeutic effects. Hofman expressed his belief in how (social) media can be used as an effective tool to communicate locally grounded emotional narratives, even if they are 'just' about excitement. For instance, in a Facebook announcement of *Rubber Duck* in Seoul, SBS News World shared:

#FlorentijnHofman, the artist who is known for his famous #RubberDuck sculpture, is posing in front of camera with yellow *Rubber Duck*. 'Korea has been through so many sad incidents lately. I wish *Rubber Duck* spreads joy and happiness to people in Korea', said Hofman at the press conference today.[15]

However, Hofman (interview) stressed that his piece is more than just about ubiquitous pleasure and fun. We should approach this artwork 'as pedagogical piece . . . It can make everyday people proud owners of their environment', he said. Hofman indicated that online sharing of his mission statement might have made publics more aware of public art's local relevance, but also of how creative online network practices could make positive global impacts.

Conclusion

This chapter aims to open up an important aspect of contemporary culture through the examination of the nexus of public art, digital and online technology, and society. Scholarship has been largely remiss in addressing people–space relationships through the lens of everyday online public art engagement beyond rather formalist interpretations of artworks in physical locations. The presented cybergeographical case study on *Rubber Duck* therefore attended to digitally networked spaces of engagement and how online roles of public art can be explored along new senses of social and spatial publicness, temporalities and uses. This article elaborated the notion of 'virtual relationality' (after Massey and Rose 2003): the assemblage of digital content from the creator, diverse media and online users expressed how public art's roles and meanings were exchanged and negotiated/challenged once mediated and hence co-created online. This linked scales of everyday offline living spaces to the translocal/global metaphysical spheres of cyberspace.

What can scholars as well as practitioners, artists, commissioners and policy makers learn from the case study presented here? First, this case has revealed an understanding of public art through socially, spatially and temporally fluid contexts among multiple (micro)publics. It has also served as a microcosm diagnostic of public art's potential online uses. Digitally mediated engagements with *Rubber Duck* implied online and offline reality as a hybrid, two-layered palimpsest for real-world encounters (in 'real life', on the Internet, in real time, offline, in synchronous or asynchronous interactions, etc.). Also, over the course of the exhibitions, user-created content produced a (lasting or potentially obsolescent) digital archive of stories and memories. This implied the amalgamation of second- and first-hand, as well as offline and online experiences/narratives published on personal or others' online user accounts. Moreover, it re-territorialised and also withered material and virtual spaces and their boundaries (see convergence culture in Jenkins 2008).

Thus, digitally mediated public art engagement is multi-layered in non-linear choreographies of space, time and society. As argued by Freeman and Sheller (2015: 6), digital public art does not only involve 'multiple modes of perception and communication[.] Beyond objects, [computer-screen-based] representations, or texts, it is about modes of activity, intensity, and coming together', which, according to them, intensify the spatial-material world, too (see De Souza e Silva 2004). Hofman (interview) remarked '*ceci n'est pas un canard*' to precisely express the 'surrealistic' nature of imaging/imagining, socialising and spatialising public art in digital culture.

As such, this study asked for a deconstruction of dominant binaries, including material/digital space, permanence/ephemerality, artist/audience, public/private and human/non-human, to understand the experience and (co-)creation of public art through interwoven offline and online media and representations (see Jenkins 2008). Through (co-)creating, sharing, commenting, curating/editing, both officials and publics showed the agency to act, self-disclose and strike their own notes, implying new media technologies' potential for informal education and cultural democratisation beyond the expert/novice binary (see Gauthier 2015, Hartley 2012). Despite digital culture's democratic changes in context, audience and status (Hartley 2012), there is still dependency on others for gaining trust and authenticity over social media (Miller 2008). This case could particularly educate public art commissioners about how meaningful engagement could be invited/explored in digitally networked spaces, which also traverse offline spaces (e.g. streets and museums), and how publics' shared participation and sense of ownership might be boosted in digital practice.

Second, this case study has demonstrated that online network practices sometimes did and sometimes did not elicit critical reflections on society and culture in regard to, for example, artistic celebration, intercultural communication, place promotion, formal or social/informal ownership, public privacy, copyright protection, (co-)authorship, preservation and archiving (see Gauthier 2015). Sometimes, online users created their own digital vignettes of *Rubber Duck*'s matter/meanings, which occasionally reproduced or undermined views by other users. Resonating with Kidd (2014), user-created content may also be interpreted as intervention in formal and mainstream journalistic practices.

Much observed e-activity was characterised by techno-human subjectification (see Goriunova 2013), revealing flattening, phatic communications (see Miller 2008). Online behaviours were largely focused on sheer social networking and fleeting, oft-entertaining cursory uses (see 'silly citizenship' in Hartley 2012), rather than on pursuing sustained dialogues, well-considered information transfer, and deep(er) awareness of one another. Online users often just scratched the geographical surface of the various exhibition spaces, as if *Rubber Duck* was cut out of its offline space and indiscriminately put online. But the travelling exhibition induced also some senses of 'belonging-in-movement'. Moreover, this case study, like the example of *Tree* discussed earlier (see also Zebracki, forthcoming), can also show how (massively) consumed artwork in physical space might reciprocally resonate with online interactions: debates, commotion, parodies, memes, etc. Heedless mass-reproduction of public art in both material and digital capacities might be, following Miller (2008), embedded within, and reinforce, probably nihilistic consequences of digital communication.

Third, this case might also be food for thought about the ethics of care for involving publics in as well as conducting research on engagements with public artwork in a digital neoliberal society (see Kidd 2014, Lodi 2014): what are the (shared) responsibilities and cosmopolitan values of online users in the (co-)creation and exchange of public art content and what do they say about respect, agony, bonding and living together? More broadly, what are sensible and accountable online research methodologies and ethics for examining user-created content? To what extent is user-created content expropriated/exploited once it is uploaded to commercially owned sites (see Hartley 2012)? What kind of publicness exists in the 'technological Panopticon' of smart city contexts and their foreseeable futures, wherein 'we ourselves produce the data that fulfil the contemporary paradigm of surveillance and control' (Lodi 2014: 283)?

Further research can move these areas further in critically exploring the potentialities and limitations regarding the use of digital media as sites of public art research and digital technologies as methodological tools to study public art engagements in the everyday life.

Acknowledgements

I am grateful to Florentijn Hofman for the interview and hospitality at his home studio. I am also thankful to David Bell, John Hartley and the anonymous referees of the original article in the *International Journal of Cultural Studies*, first published online on 9 May 2016, for their useful insights that have strengthened this work. The current version reflects some further edits. Any errors remain my own.

Notes

1 Available at: www.florentijnhofman.nl/dev/projects.php [accessed: June 2014].
2 Available at: www.florentijnhofman.nl/dev/project.php?id=192 [accessed: June 2014].
3 Available at: www.florentijnhofman.nl/dev/project.php?id=192 [accessed: June 2014].
4 These in-situ expressions and doings (through, e.g., language, social behaviours and creative acts) might, in a sense, be considered 'virtual' *avant la lettre* (Boellstorff 2008).
5 Available at: http://news.yahoo.com/video/giant-rubber-duck-explodes-taiwan-220330915.html [accessed: June 2014].

6 Translated from '悠閒享受童心的夫婦'.
7 Available at: www.nytimes.com/2015/02/15/us/museum-rules-talk-softly-and-carry-no-selfie-stick.html?_r=0 [accessed: June 2015].
8 Pauw's interview with Hofman, from which this statement and the subsequent remarks by Hofman were taken, is available at: http://pauwenwitteman.vara.nl/media/302263 [accessed: June 2014].
9 Available at: http://wantchinatimes.com/news-subclass-cnt.aspx?id=20150401000077&cid=1101 [accessed: June 2014].
10 Available at: http://knowyourmeme.com/memes/big-yellow-duck [accessed: June 2014].
11 Available at: www.citylab.com/design/2014/09/the-rubber-duck-artist-must-be-stopped/379577 [accessed: June 2014].
12 Available at: http://china.nlembassy.org/news/2013/06/original-rubber-duck-floats-to-beijing.html [accessed: January 2016].
13 Ibid.
14 Available at: www.youtube.com/watch?v=ituX4zu6N_o [accessed: June 2014].
15 Available at: www.facebook.com/SBSNEWSWORLD/posts/151632755 1938875 [accessed: June 2014].

References

Amin, A. 2002. Ethnicity and the multicultural city: living with diversity. *Environment and Planning A*, 34(6), 959–980.

Askins, K. and Pain, R. 2011. Contact zones: participation, materiality, and the messiness of interaction. *Environment and Planning D: Society and Space*, 29(5), 803–821.

Bishop, C. 2012. Digital divide. *Art Forum*, 51(1), 435–441.

Boellstorff, T. 2008. *Coming of Age in Second Life: An Anthropologist Explores the Virtually Human*. Princeton, NJ: Princeton University Press.

Brake, D. 2014. *Sharing Our Lives Online: Risks and Exposure in Social Media*. Basingstoke: Palgrave.

Crang, M., Crang, P. and May, J. 1999. *Virtual Geographies: Bodies, Space and Relations*. London: Routledge.

De Souza e Silva, A. 2004. Mobile networks and public spaces: bringing multiuser environments into the physical space. *Convergence: The International Journal of Research into New Media Technologies*, 10(2), 15–25.

Dodge, M. and Kitchin, R. 2001. *Mapping Cyberspace*. London: Routledge.

Driscoll, C. and Gregg, M. 2010. My profile: the ethics of virtual ethnography. *Emotion, Space and Society*, 3(1), 15–20.

Faulkner, S. 2016. *Poetry as Method: Reporting Research through Verse*. New York: Routledge.

Freeman, J. and Sheller, M. 2015. Editors' statement: hybrid space and digital public art. *Public Art Dialogue*, 5(1), 1–8.

Gauthier, D. 2015. Networked monumental: site, production, and distributed publics – online, and in everyday life. *Public Art Dialogue*, 5(1), 17–54.

Goriunova, O. 2013. New media idiocy. *Convergence: The International Journal of Research into New Media Technologies*, 19(2), 223–235.

Hartley, J. 2012. *Digital Futures for Cultural and Media Studies*. Oxford: Wiley-Blackwell.

Hine, C. 2000. *Virtual Ethnography*. London: Sage.

Jenkins, H. 2008. *Convergence Culture: Where Old and New Media Collide*. New York: New York University Press.

Kellerman, A. 2014. *The Internet as Second Action Space*. New York: Routledge.

Kidd, J. 2014. *Museums in the New Mediascape: Transmedia, Participation, Ethics*. Farnham: Ashgate.

Kwon, M. 2004. *One Place after Another: Site-specific Art and Locational Identity*. Cambridge, MA: MIT Press.

Lefebvre, H. 1991. *The Production of Space*. Oxford: Blackwell.

Licoppe. C, and Smoreda, Z. 2005. Are social networks technologically embedded? How networks are changing today with changes in communication technology. *Social Networks*, 27(4), 317–335.

Lodi, S. 2014. Spatial narratives in art, in *Augmented Reality Art: From an Emerging Technology to a Novel Creative Medium*, edited by V. Geroimenko. New York: Springer, 277–294.

Massey, D. and Rose, G. 2003. *Personal Views: Public Art Research Project*. Milton Keynes: Open University Press.

Miller, V. 2008. New media, networking and phatic culture. *Convergence: The International Journal of Research into New Media Technologies*, 14(4), 387–400.

Prendergast, M. 2006. Found poetry as literature review: research poems on audience and performance. *Qualitative Inquiry*, 12(2), 369–388.

Rhodes, G. 2015. Future museums now: augmented reality musings. *Public Art Dialogue*, 5(1), 59–79.

Rose, G. 2015. Rethinking the geographies of cultural 'objects' through digital technologies: interface, network and friction. *Progress in Human Geography*, 40(3), 334–351.

Song, H. 2014. All that Rubber Duck! NP: English Reporters Club (hardcopy provided by Florentijn Hofman, 11 November 2014).

Stevens, Q. and Lossau, J. 2015. Framing art and its uses in public space, in *The Uses of Art in Public Space*, edited by J. Lossau and Q. Stevens. London: Routledge, 1–16.

Warner, M. 2002. Publics and counterpublics. *Public Culture*, 14(1), 49–90.

Zebracki, M. 2012. Engaging geographies of public art: indwellers, the 'Butt Plug Gnome' and their locale. *Social & Cultural Geography*, 13(7), 735–758.

Zebracki, M. Forthcoming. Queerying public art in digitally networked space. *ACME: An International Journal for Critical Geographies*.

Afterword

13 An artist-geographer's lens

Andrew Gorman-Murray

When Martin and Joni invited me to write an afterword for their collection on public art encounters, I enthusiastically agreed. But after reading through the chapters, I started to regret my decision to agree so readily. My regret was not because of disinterest in the material or a perceived lack in the content. On the contrary, I felt overwhelmed by the incisive, high-quality scholarship of the chapters. That is, my regret reflected my own fears, insecurities and vulnerabilities. I felt out of my depth. In my scholarship, I have not directly researched or written about public art, although I have certainly skirted around related topics, such as visual culture (Gorman-Murray 2011), interior design (Gorman-Murray 2014a), museums and heritage (Gorman-Murray 2008) and creative geographies (Gorman-Murray 2016). Yet public art per se has not been part of my research agenda, and I worried about what I could sensibly say about this intelligent, coherent set of chapters on public art, space and identity.

My observations, then, are informed by a dual positioning: as well as being a social and cultural geographer, I am a practising artist – for the purpose of this afterword, let's say an 'artist-geographer' – and I draw on these combined fields of practice, knowledge and experience to make some speculative comments in response to this collection. These geographic and aesthetic positions are very much entwined. As a brief explanation, my training in social and cultural geography influences and impels my art practice, which seeks to explore the multivalent relationships between identity, place and environment. I describe my art practice as 'expanded photography', and I work with the material nature of image making, printing and installation in order to reflect the relationship between spaces, bodies and images as material objects (Gorman-Murray 2014b, 2015, 2016, forthcoming; Gorman-Murray and Brickell forthcoming).

In this afterword – being, as it is, an 'after' word – I do not want to intervene directly in the themes of the collection in the way they have been structured by Martin, Joni and the contributors. I think this ground is already well covered in the preceding pages. The sub-titled concepts of art, space and identity undergird the chapters, and the organising framework cogently captures and conveys the themes of power, affect and super-diversity as they map onto and manifest the entwining of art, space and identity. The contributors themselves examine a range of places, media, communities, subjectivities and politics that are constituted, explored or deployed by public art, and questions of audience, participation, application and evaluation are interrogated. Using my positioning as an 'artist-geographer' to respond to this comprehensive range of issues, I want to pick up on three 'alternative' themes that are introduced more or less tangentially in the collection but that I think will deserve fuller attention in future scholarship on the geographies of public art (indeed, all art practice). These themes might be briefly denoted as *mobile digital* public art, *rural* public art and *multiscalar* public art.

To be certain, the role of the digital, especially mobile digital media, in public art is probably the least tangential and most obvious of these alternative themes. Some contributors examine works of public art that either incorporate digital media in their form, production and placement (e.g. chapters by Martha Radice and Brenden Harvey; Harriet Hawkins and Ruth Catlow) or in the generation of audiences and critiques (e.g. chapters by Martha Radice and Brenden Harvey; Martin Zebracki). Notably, the role of location-based mobile digital media (involving mobile devices, mobile apps, geo-location, social networking, texting) is critically discussed in relation to newer modes of public art creation and engagement. Given the rise of digital geographies in the past decade (Ash, Kitchin and Leszczynski forthcoming), the proliferation of mobile devices and the imperative for everyday lives to be lived 'online' as well as 'offline' (Pfaff 2010, de Souza e Silva 2013), and thus the growing interpenetration of virtual and material worlds (indeed, it is becoming redundant to speak of a division between these – see Kinsley 2014), it seems to me that mobile digital media will become increasingly important to public art, both in the making of works and in public knowledge of and participation in art worlds. How might these new media continue to impel art's 'expanded fields'?

There are several pivot points that leverage the greater development of mobile digital public art. Mobile digital technologies may allow public art to be more mobile in mediated forms, as demonstrated by the works discussed by Martha Radice and Brenden Harvey (*Time Transit*, Kim Morgan, 2006) and Harriet Hawkins and Ruth Catlow (*Play Your Place*, Ruth Catlow and Mary Flanagan, 2013 forward), in which the art is digitally mediated and audiences engage with its digital elements as much as (if not more than) its material forms. Alongside its use as a component – a 'material' – within public artworks, mobile digital media might be increasingly used behind the scenes in the production of art, enabling collaborating artists to communicate ideas with each other as well as with supporting institutions and organisations. Furthermore, Martha Radice and Brenden Harvey, Harriet Hawkins and Ruth Catlow, and Martin Zebracki all demonstrate the increasing role that mobile digital media will have in generating audiences, enabling public participation and voicing critique. It seems to me that the diverse use of mobile digital media in public art will almost certainly continue to grow, and more sustained and focused investigation of its possibilities (and problems) will be a useful direction for scholarship.

The second theme I would like to see future scholarship explore is rural public art. Public art certainly features in rural as well as urban spaces, but is arguably more closely associated with the urban imaginary, with the higher population density of cities eliciting the potential for spectacle and public participation (Johnson 2009). This might be juxtaposed to 'land art' or 'earth art', a practice and movement in which the landscape is inextricably linked with and becomes part of the artwork, and which is perhaps more aligned with the rural imaginary (Shirley 2016). Indeed, most of the examples presented in this collection focus on urban public art, encompassing cities such as Amsterdam, Birmingham, Budapest, Regina, Melbourne, Singapore, Toronto and Vancouver. The exception is Pauline Guinard's discussion of the *Border Farm* project (Thenjiwe Nkosi, 2009 forward) located at Maroi (a farm) and Musina (a small town) in South Africa, on the country's border with Zimbabwe.

Pauline Guinard draws attention to themes related to rural mobilities and economies that underpin Nkosi's work, and thus highlights a relative lack of focused exploration of public art about or located in the rural. Given what seems to be a global concern about the plight of rural populations in the face of worldwide urbanisation (Woods 2011), I suggest that another useful field for further research is the role and meaning of public art in rural communities. For me, this also responds to recent geographical work from Australia that has examined the benefits of rural festivals and creativity, which have been used to sustain townships, reinforce place-attachment, foster wellbeing and bolster local economies, while offering pointed political claims about the needs of rural communities in relation to urban-facing policies from governments (Gibson and Connell 2011, 2012).

The third and final theme I want to attend to concerns the multiscalarity of public art. This is not so much a call to produce public artwork that is multiscalar (in the way we might think of making digital or rural public artwork), but rather a suggestion to explore and make explicit the way public artworks transcend multiple scales in their politics and power, from the body to the globe (Herrod 2011). While public artworks are site-specific and localised – in the cases here, often in urban sites – their messages, meanings and affects reach across various scales and link them together. For instance, thinking of Jason Luger's discussion of the *Awaken the Dragon* festival in Singapore, we can see the significance of the art project taking different registers at different scales: affectively, through the embodied practices of artists and audiences; locally, through the call to save the last Dragon Kiln on the island; nationally, in terms of sustaining a piece of cultural heritage and Singaporean 'identity'; and globally, in terms of the government's desire to craft Singapore as a global creative city and arts hub through policy and promotion.

Jason Luger shows that these multiscalar meanings have different political intentions and relationships to circuits of social power in Singapore. Some are aligned; some are in tension. But the interesting point, I think, is that we can use the public artwork to explore how different scalar politics are assembled around (and through) it, and thus reveal where productive sticking points – or points of agonistic debate (Mouffe 2007) – emerge. In a similar vein, Martin Zebracki shows how the multi-sited iterations of Florentijn Hofman's *Rubber Duck* (2012 forward) – including installations in Amsterdam, Baku, Beijing, Pittsburgh, Sydney and elsewhere – bring different permutations of personal, local, national and global politics into dialogue through the use of mobile digital technologies that foster a multiscalar public engagement beyond each specific site, across the virtual/material divide. Again, we can use public art to examine the assemblage of a multiscalar politics. In this way, public art connects with identity, politics and power in multivalent ways that extend beyond the intentions of artists, local populations or local states, and are revealed through investigation of the affects of the artwork for different people at multiple scales (see also the chapters by Paul Clements; Gwen McGregor; Aaron Franks, Jess Masuda and Audrey Kobayashi).

There is, of course, more that could be said, and more themes to be advanced in future work, such as the precarious and tenuous borders between public and private engagement by practitioners and participants in public art (e.g. chapters by Martha Radice and Brenden Harvey; Paul Clements; Harriet Hawkins and Ruth Catlow; Candice Boyd) and between social inclusions and exclusions in relation to community-facing public art projects (e.g. chapters by Cameron Cartiere and Ashley Guindon; Aaron Franks, Jess Masuda and Audrey Kobayashi; Saskia Warren and Stephen Forcer). But working through this afterword from the lens of an artist-geographer, I chose to highlight three alternative themes for spatially sensitive art practice and critique that emerged from my reading of the collection and that could be generative for future work. These themes, located at the intersection of geography and art practice, are mobile digital public art, rural public art and multiscalar public art. I look forward to future art practice and geographical research (and especially their intersections) that builds on and interrogates these arenas. The bourgeoning work of experimental and creative geographies might be particularly incisive here (Enigbokan and Patchett 2012; Hawkins 2013a, 2015; Hunt 2014; Last 2012; Marston and De Leeuw 2013; Paglen 2008). For as Harriet Hawkins (2013b: 53) urged in her discussion of the intersecting and expanding fields of geography and art, geographers may benefit by 'taking seriously art as a mode of critical exploration' and embracing 'practices normally deemed "artistic", or more broadly "creative", within their own methodological repertoires'.

References

Ash, J., Kitchin, R. and Leszczynski, A. Forthcoming. Digital turn, digital geographies? *Progress in Human Geography*.

de Souza e Silva, A. 2013. Location-aware mobile technologies: historical, social and spatial approaches. *Mobile Media and Communication*, 1(1): 116–121.

Enigbokan, A. and Patchett, M. 2012. Speaking with spectres: experimental geographies in practice. *Cultural Geographies*, 19(4), 535–546.

Gibson, C. and Connell, J. 2012. *Music Festivals and Regional Development in Australia*. Farnham: Ashgate.

Gibson, C. and Connell, J. (eds). 2011. *Festival Places: Revitalising Rural Australia*. Bristol: Channel View.

Gorman-Murray, A. 2008. So, *where* is queer? A critical geography of queer exhibitions in Australia. *Museums and Social Issues*, 3(1): 67–80.

Gorman-Murray, A. 2011. 'This is disco-wonderland!' Gender, sexuality and the limits of gay domesticity on *The Block*. *Social and Cultural Geography*, 12(5): 435–453.

Gorman-Murray, A. 2014a. Materiality, masculinity and the home: men and interior design, in *Masculinities and Place*, edited by A. Gorman-Murray and P. Hopkins. Farnham: Ashgate, 209–226.

Gorman-Murray, A. with Brickell, C. and de Jong, A. 2014b. *Over the Ditch*, in *On Islands Creative Collaboration Festival*, Eramboo Artist Environment, Terrey Hills, NSW, 15–30 November, catalogue by S. Griffith (ISBN 9781634430210).

Gorman-Murray, A. 2015. *Material Conditions in the Post-Human City*, in *Thresholds*, group exhibition at Ivan Dougherty Gallery, Sydney, 17–20 June 2015, catalogue by J. Santamaria, Lyra and R. Spark (ISBN 9780987138255).

Gorman-Murray, A. 2016. *Thrown-togetherness*, 2015: exegesis, in *Housing and Home Unbound: Intersections in Economics, Environment and Politics in Australia*, edited by L. Crabtree, N. Cook and A. Davison. Routledge: London, pp. 18–21, 94–97, 168–171, 232–235.

Gorman-Murray, A. Forthcoming. *Thrown-togetherness*: queering the interior in visual perspectives. In *Queering the Interior*, edited by A. Gorman-Murray and M. Cook. Bloomsbury: London.

Gorman-Murray, A. and Brickell, C. Forthcoming. *Over the Ditch*: Queer mobilities at the nexus of art, geography and history. *ACME: An E-Journal of Critical Geographies*.

Hawkins, H. 2013a. *For Creative Geographies: Geography, Visual Art and the Making of Worlds*. London: Routledge.

Hawkins, H. 2013b. Geography and art. An expanding field: site, the body and practice. *Progress in Human Geography*, 37(1): 52–71.

Hawkins, H. 2015. Creative geographies: knowing, representing, intervening. *Cultural Geographies*, 22(2): 247–268.

Herrod, A. 2011. *Scale*. Routledge: London.

Hunt, M. 2014. Urban photography/cultural geography: spaces, objects, events. *Geography Compass*, 8/3: 151–168.

Johnson, L. 2009. *Cultural Capitals: Revaluing the Arts, Remaking Urban Spaces*. London: Routledge.

Kinsley, S. 2014 The matter of 'virtual' geographies. *Progress in Human Geography*, 38(3): 364–384.

Last, A. 2012. Experimental geographies. *Geography Compass*, 6/12: 706–724.

Marston, S. and De Leeuw, S. 2013. Creativity and geography: toward a politicized intervention. *Geographical Review*, 103(2): ii–xxvi.

Mouffe, C. 2007. Artistic activism and agonistic spaces. *Art and Research: A Journal of Ideas, Contexts and Methods*, 1(2). Available at: www.artandresearch.org.uk/v1n2/mouffe.html [accessed: 23 May 2017].

Paglen, T. 2008. Experimental geography: from cultural production to the production of space. In *Experimental Geography*, edited by N. Thompson. New York: Melville House, 27–33.

Pfaff, J. 2010. Mobile phone geographies. *Geography Compass*, 4(10): 1433–1447.

Shirley, R. 2016. *Rural Modernity, Everyday Life and Visual Culture*. London: Routledge.

Woods, M. 2011. *Rural*. London: Routledge.

Appendix

Response piece: everyday lived experience

Zodwa Nyoni

A poem constructed through the found poetry method (e.g. Faulkner 2016; Prendergast 2006), using 'A cybergeography of public art encounter: the case of *Rubber Duck*' (Chapter 12) by Martin Zebracki.

Use cyberspace to capture
temporary people and things.
Turn hybrid between old and new,
Merge in space and time.

Human to cultural objects.
This is a case study.

The world reconfigured as urban installations.
The central mission,
Bring different people and cultures together,
Throw light.

Place virtual relationality
under critique in everyday life.
Starting from the offline world.

A concern of time, ephemerality.
The very essence of meaning,
Negotiation between identities,
Size, colour, shape, position, permanence.

Not all people engage with the same gamut of properties.
They pursue these properties to different degrees
and in different ways.

I extend the notion, mediation,
Repurposing, challenging properties
and roles in networked spaces.

Focus on social reconstructions of public communities
Twitter, Facebook, Youtube, forums
and online news platforms.
Digital culture.

Social media have substantially affected
and intensified quotidian life.
Mass engagement is layered
Over people's real world experiences.
Technologies are producing,
different human experience of dwelling in the world.
New articulations
Of near and far
Present and absent
Body and technology
Self and environment.

Multiple in-situ users,
within single environments.

Such messiness of space.
Dispersed potential,
Invisible identities,
Uploading, downloading
and streaming.
Augmenting one another
and deconstructing our existence.
Blurred boundaries.

Cultural objects for citizens.
Shift from people to digital spheres.

Second hand art,
Somewhere in between
Entity
Or reality.

Note on the provenance of the poem

Martin Zebracki was selected for Leeds Creative Labs within the remit of the 2017 Edition of DARE, the partnership between Opera North and the University of Leeds. He was matched with Tyrrell Jones, director of Knaïve Theatre, a Leeds-based performing arts partnership that aims to connect, challenge and empower public audiences. On 5 July 2017, Zebracki and Jones performed the showcase entitled *The Ecology of Digital Life: Dear Humans and Cyborgs* in further collaboration with the poet and playwright Zodwa Nyoni, the dancer Roseanna Anderson, and the composer Robert Bentall. This live-streamed showcase formed a collective response to Zebracki's recent work in the digitally mediated geographies of public art, including his contribution 'A cybergeography of public art encounter: the case of *Rubber Duck*' (Chapter 12) – the above poem ensued as one of the response pieces from the collaboration.

References

Faulkner, S. 2016. *Poetry as Method: Reporting Research through Verse*. New York: Routledge.

Prendergast, M. 2006. Found poetry as literature review: research poems on audience and performance. *Qualitative Inquiry*, 12(2), 369–388.

Index

Page numbers in **bold** refer to illustrations.

www.ingramcontent.com/pod-product-compliance
Ingram Content Group UK Ltd.
Pitfield, Milton Keynes, MK11 3LW, UK
UKHW020010200926
475702UK00014B/1804
9780367362102